CALVERT MATH

Calvert Math is based upon a previously published textbook series. Calvert School has customized the textbooks using the mathematical principles developed by the original authors. Calvert School wishes to thank the authors for their cooperation. They are:

Audrey V. Buffington
Mathematics Teacher
Wayland Public Schools
Wayland, Massachusetts

Alice R. Garr
Mathematics Department Chairperson
Herricks Middle School
Albertson, New York

Jay Graening
Professor of Mathematics
  and Secondary Education
University of Arkansas
Fayetteville, Arkansas

Philip P. Halloran
Professor, Mathematical Sciences
Central Connecticut State University
New Britain, Connecticut

Michael Mahaffey
Associate Professor,
  Mathematics Education
University of Georgia
Athens, Georgia

Mary A. O'Neal
Mathematics Laboratory Teacher
Brentwood Unified Science
  Magnet School
Los Angeles, California

John H. Stoeckinger
Mathematics Department Chairperson
Carmel High School
Carmel, Indiana

Glen Vannatta
Former Mathematics Supervisor
Special Mathematics Consultant
Indianapolis Public Schools
Indianapolis, Indiana

ISBN-13: 978-1-888287-57-8

Copyright © 2008 by Calvert School, Inc.
All rights reserved. No part of this book may be reproduced or transmitted in any form or by any means, electronic or mechanical, including photocopying, recording, or by any information storage and retrieval system, without permission in writing from the publisher.

1 2 3 4 5 6 7 8 9 10   12 11 10 09 08

SENIOR CONSULTANT/PROJECT COORDINATOR
Jessie C. Sweeley, Manager of Curriculum Development

PROJECT FACILITATOR/LEAD RESEARCHER
Nicole M. Henry, Math Curriculum Specialist

FIELD TEST TEACHERS
The following Calvert teachers used the Calvert Math prototype in their classrooms and added materials and teaching techniques based on Calvert philosophy and methodology.

Andrew S. Bowers	Brian J. Mascuch	Patricia G. Scott
James O. Coady Jr.	John M. McLaughlin	E. Michael Shawen
Roman A. Doss	Mary Ellen Nessler	Ada M. Stankard
Shannon C. Frederick	Margaret B. Nicolls	Virginia P.B. White
Julia T. Holt	Michael E. Paul	Jennifer A. Yapsuga
Willie T. Little III	Diane E. Proctor	Andrea L. Zavitz

CALVERT EDUCATION SERVICES (CES) CONSULTANTS
The following CES staff worked with the Day School faculty in developing the mathematical principles for Calvert Math.

Eileen M. Byrnes	Christina L. Mohs	Mary-Louise Stenchly
Nicole M. Henry	Kelly W. Painter	Ruth W. Williams
Linda D. Hummel		

REVISION AUTHORS
The following Day School faculty and CES staff contributed to the correlation of the math materials to national standards and to the authoring of revisions based on the correlations.

Nicole M. Henry	Willie T. Little III	Mary Ethel Vinje
Julia T. Holt	Mary Ellen Nessler	Jennifer A. Yapsuga
Barbara B. Kirby		

COPY EDITORS
Bernadette Burger, Senior Editor
Sarah E. Hedges   Mary Pfeiffer
Maria R. Kerner   Megan L. Snyder

GRAPHIC DESIGNERS
Vickie M. Johnson, Senior Designer
Vanessa Ann Panzarino

To the Student
This book was made for you. It will teach you many
new things about mathematics. Your math book is
a tool to help you discover and master the skills
you will need to become powerful in math.

About the Art in this Book
Calvert homeschooling students from all over the world
and Calvert Day School students have contributed
their original art for this book. We hope you enjoy
looking at their drawings as you study mathematics.

Contents

Diagnosing Readiness

Diagnostic Skills Pretest xiv
DIAGNOSTIC SKILLS
1 Comparing and Ordering Decimals xvi
2 Adding and Subtracting Decimals.............xvii
3 Multiplying and Dividing Decimals xviii
4 Divisibility Patterns xix
5 Prime Factors xx
6 Greatest Common Factor xxi
7 Least Common Multiple xxii
8 Plotting Points on a Coordinate Plane xxiii
9 Converting Units within the Metric
   System... xxiv
10 Converting Measurements within
    the Customary System xxvi
Diagnostic Skills Posttest xxviii

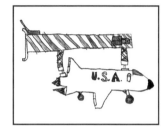

1 Integers and Algebra

1.1 Order of Operations................................2
　　MIND BUILDER: Patterns 3
1.2 Variables 4
1.3 Expressions and Equations 6
1.4 Properties 8
PROBLEM SOLVING
　　Guess Again...10
CUMULATIVE REVIEW..11
1.5 Integers and Absolute Value 12
1.6 Adding and Subtracting Integers 14
1.7 Multiplying and Dividing Integers 16
　　Mid-Chapter Review 18
PROBLEM SOLVING
　　Which One is Different?..........................19
1.8 Inverse and Distributive Properties 20
　　MIND BUILDER: Riddle.................................21
1.9 Solving One-Step Equations...................22
1.10 Solving Two-Step Equations.................24
　　MIND BUILDER: Find the Equation 25
1.11 Problem-Solving Strategy:
　　Write an Equation 26

Chapter 1 Review...28
Chapter 1 Test 30
CHANGE OF PACE
　　Modular Arithmetic 31
Cumulative Test 32

Contents　v

2 Fractions and Exponents

2.1 Fractions and Decimals 34
2.2 Comparing and Ordering Rational Numbers 36
2.3 Adding and Subtracting Fractions and Mixed Numbers........................38
   MIND BUILDER: Adding Algebraic Fractions..................................40
PROBLEM SOLVING
   Whodunit?..................................41
2.4 Multiplying and Dividing Fractions and Mixed Numbers........................42
   MIND BUILDER: Dividing Algebraic Fractions..................................44
CUMULATIVE REVIEW..........................45
2.5 Solving Equations with Fractions and Mixed Numbers........................46
2.6 Problem-Solving Strategy: Solve a Simpler Problem 48
   Mid-Chapter Review 49
2.7 Exponents...50
2.8 Properties of Exponents 52
2.9 Negative and Zero Exponents................54
2.10 Scientific Notation 56
   MIND BUILDER: Decimals in Expanded Form..................................57

Chapter 2 Review..58
Chapter 2 Test 60
CHANGE OF PACE
   History......................................61
Cumulative Test 62

3 Real Numbers

3.1 Squares and Square Roots........................64
3.2 Real Numbers 66
3.3 The Pythagorean Theorem........................68
   Mid-Chapter Review 69
3.4 Using the Pythagorean Theorem.............70
PROBLEM SOLVING
   What Color is It?..72
CUMULATIVE REVIEW....................................73
3.5 30°–60° Right Triangles 74
3.6 45°–45° Right Triangles 76
   MIND BUILDER: Pythagorean Triples 77
3.7 Problem-Solving Strategy: Using Logical Reasoning 78
   MIND BUILDER: Puzzle..............................79

Chapter 3 Review..80
Chapter 3 Test 82
CHANGE OF PACE
   Logic 83
Cumulative Test 84

4 Ratios and Proportions

4.1 Ratios and Rates 86
4.2 Rate of Change 88
4.3 Slope 90
CUMULATIVE REVIEW 93
4.4 Conversions Using Dimensional
    Analysis 94
4.5 Proportions 96
4.6 Problem-Solving Strategy:
    Using a Proportion 98
    Mid-Chapter Review 99
4.7 Similar Figures 100
    MIND BUILDER: Using a Stadia 102
PROBLEM SOLVING
    The Problem Problem 103
4.8 Dilations 104
4.9 Scale Drawings 106
4.10 Problem-Solving Application:
    Using Indirect Measurement 108
4.11 Tangent, Sine, and Cosine Ratios 110
4.12 Problem-Solving Application:
    Using Tangent, Sine, and Cosine 112

Chapter 4 Review 114
Chapter 4 Test 116
CHANGE OF PACE
    Angle of Elevation 117
Cumulative Test 118

5 Percents

5.1 Fractions, Decimals, and Percents 120
5.2 Finding Percents Mentally 122
5.3 Percents and Proportions 124
    MIND BUILDER: Mental Math 125
5.4 Percent Equations 126
    Mid-Chapter Review 127
PROBLEM SOLVING
    Super Fly 128
CUMULATIVE REVIEW 129
5.5 Percent of Change 130
5.6 Markup, Discount, and Sales Tax 132
5.7 Simple and Compound Interest 134
5.8 Problem-Solving Strategy:
    Guess and Check 136

Chapter 5 Review 138
Chapter 5 Test 140
CHANGE OF PACE
    Egyptian Mathematics 141
Cumulative Test 142

6 Polygons and Transformations

6.1 Angle Relationships	144
6.2 Parallel Lines and Angles	146
MIND BUILDER: Angles	148
CUMULATIVE REVIEW	149
6.3 Constructions: Line Segments and Angles	150
6.4 Constructions: Bisectors of Line Segments and Angles	152
6.5 Constructions: Perpendiculars and Parallels	154
6.6 Problem-Solving Application: Constructing Golden Rectangles	156
Mid-Chapter Review	157
6.7 Triangles and Quadrilaterals	158
6.8 Polygons and Angles	160
6.9 Congruent Figures	162
6.10 Translations and Rotations	164
MIND BUILDER: Logical Thinking I	166
PROBLEM SOLVING	
Tricky Cuts	167
6.11 Reflections and Symmetry	168
6.12 Problem-Solving Strategy: Identifying Necessary Facts	170
Chapter 6 Review	172
Chapter 6 Test	174
CHANGE OF PACE	
Tessellations	175
Cumulative Test	176

7 Measuring Area and Volume

7.1 Area	178
PROBLEM SOLVING	
Carpet the Library	181
7.2 Circles	182
CUMULATIVE REVIEW	185
7.3 Three-Dimensional Figures	186
7.4 Volume of Prisms and Cylinders	188
Mid-Chapter Review	189
7.5 Volume of Pyramids and Cones	190
7.6 Surface Area of Prisms and Pyramids	192
MIND BUILDER: Logical Thinking II	193
7.7 Surface Area of Cylinders and Cones	194
7.8 Problem-Solving Strategy: Using a Diagram	196
Chapter 7 Review	198
Chapter 7 Test	200
CHANGE OF PACE	
Spheres	201
Cumulative Test	202

8 Data and Statistics

8.1 Samples and Surveys............................204
8.2 Circle Graphs 206
8.3 Central Tendency.................................208
8.4 Stem-and-Leaf Plots 210
8.5 Measures of Variation..........................212
    MIND BUILDER: Mean Variation 213
PROBLEM SOLVING
    Pentominoes......................................214
CUMULATIVE REVIEW....................................215
8.6 Box-and-Whisker Plots 216
    Mid-Chapter Review 217
8.7 Choosing an Appropriate Display..........218
8.8 Misleading Graphs 220
8.9 Scatter Plots..222
    MIND BUILDER: Percentile...........................223
8.10 Problem-Solving Strategy:
    Looking for a Pattern 224

Chapter 8 Review...226
Chapter 8 Test 228
CHANGE OF PACE
    Cumulative Frequency Histograms...........229
Cumulative Test 230

9 Probability

9.1 Probability 232
9.2 Counting Outcomes 234
9.3 Permutations 236
9.4 Combinations 238
9.5 Pascal's Triangle 240
    Mid-Chapter Review 241
9.6 Theoretical and Experimental
    Probability 242
PROBLEM SOLVING
    Name That Cube 244
CUMULATIVE REVIEW....................................245
9.7 Independent and Dependent Events 246
9.8 Odds 248
9.9 Problem-Solving Strategy:
    Acting It Out 250
    MIND BUILDER: Logical Thinking III 251

Chapter 9 Review...252
Chapter 9 Test 254
CHANGE OF PACE
    Is This Game Fair?..................................255
Cumulative Test 256

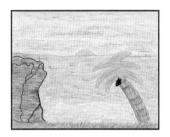

10 More Equations and Inequalities

10.1 Writing Two-Step Equations 258
10.2 Simplifying Algebraic Expressions........260
10.3 Solving Multi-Step Equations 262
    Mid-Chapter Review 263
10.4 Solving Equations with Variables
    on Both Sides 264
PROBLEM SOLVING
    The Container Problem 266
CUMULATIVE REVIEW......................................267
10.5 Solving Inequalities by Addition
    and Subtraction 268
10.6 Solving Inequalities by
    Multiplication and Division 270
10.7 Solving Two-Step Inequalities..............272
10.8 Problem-Solving Application:
    Using Inequalities 274

Chapter 10 Review......................................276
Chapter 10 Test 278
CHANGE OF PACE
    Computers: Binary System......................279
Cumulative Test 280

11 Linear Functions

11.1 Sequences 282
11.2 Functions..284
11.3 Graphing Linear Functions 286
11.4 Graphing Two Equations.....................288
    Mid-Chapter Review 289
PROBLEM SOLVING
    The Pyramid of Basketballs.....................290
CUMULATIVE REVIEW....................................291
11.5 Slope of a Line 292
11.6 Slope-Intercept Form 294
11.7 Graphing Inequalities..........................296
    MIND BUILDER: Second-Degree Equations..297
11.8 Problem-Solving Strategy:
    Using Graphs..298

Chapter 11 Review......................................300
Chapter 11 Test 302
CHANGE OF PACE
    Rule of 78 303
Cumulative Test 304

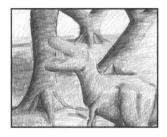

12 Algebraic Relationships

12.1 Nonlinear Functions 306
12.2 Quadratic Functions 308
12.3 Polynomials 310
  Mid-Chapter Review 311
PROBLEM SOLVING
  Undo the Magic 312
CUMULATIVE REVIEW 313
12.4 Adding and Subtracting Polynomials .. 314
  MIND BUILDER: Magic Squares 315
12.5 Multiplying Polynomials 316
12.6 Problem-Solving Strategy:
  Working Backward 318

Chapter 12 Review 320
Chapter 12 Test 322
CHANGE OF PACE
  Cubic Functions 323
Cumulative Test 324

Appendix 325
Glossary 329
Index 339

Diagnosing Readiness

In this preliminary chapter, you will take a diagnostic skills pretest, practice skills that you need to work on, and take a diagnostic skills posttest. This preliminary chapter will assure you have the skills necessary to begin Chapter 1.

Diagnostic Skills Pretest

Complete the diagnostic skills pretest below. If you get one or more problems in a section incorrect, please refer to the section number in parenthesis for review. After reviewing this section, take the diagnostic skills posttest at the end of the chapter.

Replace each ■ with <, >, or = to make a true statement. (Section 1)

1. 6.35 ■ 6.57
2. 7.82 ■ 7.50
3. 12.72 ■ 12.1
4. 0.46 ■ 0.42
5. 3.640 ■ 3.640
6. 5.38 ■ 5.38
7. 2.2 ■ 2.62
8. 6.45 ■ 6.5

Order each set of decimals from least to greatest. (Section 1)

9. 8.80, 8.35, 8.70
10. 12.7, 12.86, 12.77, 12.8

Find each sum or difference. (Section 2)

11. 23.6 + 41.35
12. 2.36 + 1.42
13. 43.84 − 28.548
14. 47.94 − 5.43
15. 16.8 − 14.2
16. 18.4 − 5.2
17. 13.204 + 80.271
18. 16.55 + 52.30
19. 3.20 − 1.06
20. 4.2 + 3.5 + 7.9
21. 6.76 − 5.04
22. 32.54 + 10.01

Find each product or quotient. (Section 3)

23. 59 • 0.4
24. 0.24 ÷ 0.125
25. 16.5 • 0.8
26. 70.8 • 2.6
27. 5.016 ÷ 3.8
28. 72 ÷ 0.08
29. 4.8 ÷ 6
30. 22.6 • 7.4
31. 2.108 ÷ 3.4
32. 6.27 • 4.2
33. 1.84 ÷ 0.8
34. 3.47 • 0.5

Determine whether each number is divisible by 2, 3, 4, 5, 6, 9, or 10. (Section 4)

35. 42
36. 85
37. 168
38. 352
39. 1,048
40. 63
41. 346
42. 765

Determine whether each number is *prime* or *composite*. Write the prime factorization of each composite number. (Section 5)

43. 14
44. 25
45. 36
46. 100
47. 27
48. 42
49. 144
50. 28

Find the greatest common factor of each set of numbers. (Section 6)

51. 5 and 14
52. 21 and 7
53. 6 and 27
54. 14 and 20
55. 20 and 25
56. 16, 24, and 40

Find the least common multiple of each set of numbers. (Section 7)

57. 3 and 2
58. 8 and 3
59. 12 and 4
60. 10 and 4
61. 7 and 5
62. 3, 6, and 12

Name the ordered pair for the coordinates of each point on the coordinate plane. (Section 8)

63. A
64. B
65. C
66. D
67. E
68. F

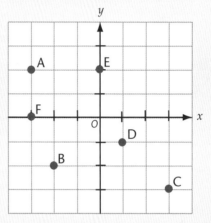

Use graph paper to graph each point on the same coordinate plane. (Section 8)

69. M(3, 5)
70. H(-2, 6)
71. N(4, -1)
72. T(0, -5)

Complete. (Section 9)

73. 20 kg = _____ g
74. 4 L = _____ mL
75. 400 cm = _____ m
76. 0.4 km = _____ m
77. 100 g = _____ kg
78. 3,000 mL = _____ L

Complete. (Section 10)

79. 9 lb = _____ oz
80. 8 pt = _____ qt
81. 4 T = _____ lb
82. 4.5 ft = _____ in.
83. 9 c = _____ fl oz
84. 4 mi = _____ yd

1 Comparing and Ordering Decimals

To compare decimals:

- First attach a zero, if needed, so that each decimal has the same number of decimal places.
- Start at the left and compare the digits in each place-value position.
- Compare as with whole numbers.
- At the first decimal place where the digits are different, you can determine which number is greater and which is lesser.

Examples

A. Compare 5.35 and 5.19.

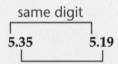

In the tenths place, 3 > 1.

5.35 > 5.19

B. Order 9.8, 9.32, and 9.65 from least to greatest.

9.32 is less than both 9.65 and 9.8.

9.65 is less than 9.8.

In order from least to greatest, the numbers are 9.32, 9.65, and 9.8.

Exercises

Replace each ■ with <, >, or = to make a true statement.

1. 3.5 ■ 3.17
2. 4.49 ■ 4.5
3. 7.82 ■ 7.60
4. 12.68 ■ 12.1
5. 6.14 ■ 6.1
6. 12.75 ■ 12.8
7. 6.150 ■ 6.27
8. 14.76 ■ 14.6
9. 9.8 ■ 9.88
10. 6.100 ■ 6.240
11. 3.750 ■ 3.750
12. 15.72 ■ 15.8

Order each set of decimals from least to greatest.

13. 2.1, 2.05, 2.25
14. 8.46, 8.81, 8.78
15. 45.9, 45.35, 45.68
16. 31.7, 31.86, 31.77, 31.18
17. 47.24, 47.25, 47.12, 47.4
18. 58.80, 58.35, 58.70
19. 0.963, 0.391, 0.299, 0.309

2 Adding and Subtracting Decimals

To add and subtract decimals, write the numbers in a column and line up the decimal points. Add or subtract the place-value positions.

Examples

A. Add. 14.3 + 6.9

```
  14.3     Line up the decimal points.
+  6.9     Add.
------
  21.2
```

B. Subtract. 82.35 − 27.14

```
  82.35    Line up the decimal points.
− 27.14    Subtract.
------
  55.21
```

You may need to place zeros at the end of the decimals to help align the columns. Then add or subtract the place-value positions.

C. Subtract. 9 − 1.85

```
  9.00     Rewrite 9 as 9.00 to align the columns.
− 1.85     Subtract.
------
  7.15
```

Exercises

Find each sum or difference.

1. 43.21 + 35.90
2. 7.96 − 5.25
3. 29.145 + 9.852
4. 16 − 1.92

5. 4.6 − 3.5
6. 3.8 + 5.1
7. 6.50 + 15.92
8. 5.54 − 2.3
9. 34.2 − 12.4
10. 47.8 + 34.86
11. 56.77 − 12.45
12. 7.89 − 4.39
13. 15.6 + 20.71
14. 4.2 − 2.01
15. 3.726 + 9.294
16. 1.8 − 1.54
17. 40.6 + 4.06
18. 0.7 + 1.28
19. 9.38 − 3.852
20. 12 − 3.74
21. 3.85 + 27.49
22. 38 − 0.64
23. 3.98 + 7.38 + 2.42
24. 0.4 + 2.85 + 0.184
25. 28.66 + 9.061 + 15.55

26. Find the sum of 78.62 and 39.64.
27. What is the difference between 14.9 and 12.72?
28. Lacey bought a pizza for $12.65, soda for $2.09, and salad for $6.84. What is the total?

3 Multiplying and Dividing Decimals

To multiply decimals, multiply as with whole numbers. Then add the decimal places in the factors together. Place the same number of decimal places in the product, counting from right to left.

Examples

A. Find the product. 11.3×3.2

```
   11.3      1 decimal place
 × 3.2       1 decimal place
   226
  339
 36.16       2 decimal places
```

The product is 36.16.

B. Find the product. 8.45×0.7

```
   8.45      2 decimal places
 × 0.7       1 decimal place
  5.915      3 decimal places
```

The product is 5.915.

To divide decimals, move the decimal point in the divisor to the right until the divisor is a whole number. Then move the decimal point in the dividend the same number of places as it was moved in the divisor. Then divide.

C. Find the quotient. $55.86 \div 9.8$

```
        5.7
  9.8)55.86
     - 490
       686
     - 686
         0
```

Move each decimal point 1 place to the right.

The quotient is 5.7.

Exercises

Find each product or quotient.

1. 9.14 • 1.3
2. 2.008 • 8.1
3. 39.78 ÷ 11.7
4. 42.66 ÷ 7.9
5. 101.92 ÷ 9.8
6. 12.38 • 3.85
7. 121.26 • 9.4
8. 4.46 • 3.8
9. 16.24 ÷ 5.8
10. 57.8 ÷ 6.8
11. 32.4 • 5.9
12. 8.492 • 1.2
13. 135.42 ÷ 12.2
14. 71.98 ÷ 11.8
15. 0.205 • 6.4
16. 17.04 ÷ 2.4
17. 68.5 • 3.4
18. 60.48 ÷ 5.6
19. 54.82 • 3.2
20. 41.28 ÷ 4.8

4 Divisibility Patterns

If a number is a factor of a given number, you can also say the given number is divisible by the factor.

The rules for divisibility are to the right.

Factor	Rule	Example
2	The ones digit is divisible by 2. The number is *even*.	2, 4, 6, 8, 10, 12, . . .
3	The sum of the digits is divisible by 3.	567 5 + 6 + 7 = 18 18 ÷ 3 = 6 Therefore, 567 is divisible by 3.
4	The number formed by the last two digits is divisible by 4.	4, 8, 12, 112 112: 12 ÷ 4 = 3
5	The ones digit is 0 or 5.	5, 10, 15, 20, 25, . . .
9	The sum of the digits is divisible by 9.	459: 4 + 5 + 9 = 18 18 ÷ 9 = 2 Therefore, 459 is divisible by 9.
10	The ones digit is 0.	10, 20, 30, 40, . . .

The rules for 6 are related to the rules for 2 and 3.

Factor	Rule	Example
6	The number is divisible by both 2 and 3.	564 2: Last digit is 4 3: 5 + 6 + 4 = 15, 15 ÷ 3 = 5

Example

Determine whether 2,124 is divisible by 2, 3, 4, 5, or 6.

2: Yes, the ones digit 4 is divisible by 2.

3: Yes, the sum of the digits 2 + 1 + 2 + 4 = 9 is divisible by 3.

4: Yes, the number formed by the last two digits 24 is divisible by 4.

5: No, the number does not end in 0 or 5.

6: Yes, the number is divisible by 2 and 3, so it is divisible by 6.

Exercises

Determine whether each number is divisible by 2, 3, 4, 5, 6, 9, or 10.

1. 54
2. 120
3. 418
4. 1,500
5. 366
6. 387
7. 101
8. 464
9. 2,585
10. 907
11. 50,700
12. 232
13. Is 4 a factor of 116?
14. Is 6 a factor of 333?

5 Prime Factors

When a whole number greater than 1 has exactly two factors, itself and 1, it is called a **prime number**. When a whole number greater than 1 has more than two factors, it is called a **composite number**. The numbers 0 and 1 are neither prime nor composite because 0 has many factors and 1 has only one factor, itself.

You can determine whether numbers are prime or composite by using the divisibility rules.

Examples

A. Is 48 prime or composite?
48 is divisible by 2, 4, 6, and 8, therefore it is composite.

B. Is 53 prime or composite?
53 is only divisible by 1 and 53, therefore it is prime.

Every composite number can also be expressed as a product of prime numbers. This is called **prime factorization**. A factor tree can be used to find the prime factorization of a number.

C. Find the prime factorization of 48.

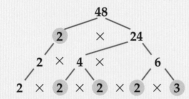 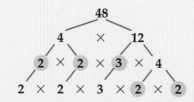 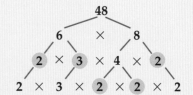

Continue to factor any number that is not prime until all factors are prime.

The prime factorization of 48 is $2 \times 2 \times 2 \times 2 \times 3$, or $2^4 \times 3$.

Exercises

Determine whether each number is *prime* or *composite*. Write the prime factorization of each composite number.

1. 12
2. 34
3. 37
4. 36
5. 16
6. 25
7. 51
8. 74
9. 20
10. 70
11. 15
12. 30
13. 72
14. 11
15. 49
16. 75
17. 225
18. 50

6 Greatest Common Factor

Groups of two or more numbers have common factors. The greatest of these is the **greatest common factor** (GCF) of the numbers. There are different ways to find the GCF.

Examples

A. Find the greatest common factor of 12 and 18.

List the factors.

factors of 12: 1, 2, 3, 4, 6, 12
factors of 18: 1, 2, 3, 6, 9, 18

Common factors of 12 and 18: 1, 2, 3, 6

The greatest common factor of 12 and 18 is 6.

You can use prime factorization.

$12 = 2 \times 2 \times 3$
$18 = 2 \times 3 \times 3$

Common prime factors of 12 and 18: 2 and 3

The GCF is 2×3, or 6.

You can also use prime factor trees to find the greatest common factor.

B. Find the greatest common factor of 15, 25, and 35.

15: 3 × 5   25: 5 × 5   35: 7 × 5

The greatest common factor of 15, 25, and 35 is 5.

Exercises

Find the greatest common factor of each set of numbers.

1. 40 and 12
2. 36 and 52
3. 27 and 54
4. 96 and 60
5. 18 and 54
6. 25 and 45
7. 44 and 55
8. 30 and 42
9. 7 and 10
10. 24, 36, and 48
11. 55, 60, and 95
12. 46, 76, and 82

7 Least Common Multiple

A multiple of a number is the product of the number and any whole number. When you learned to multiply, you learned multiples of numbers.

Examples

A. List the first six multiples of 12.

$1 \times 12 = 12$ $\qquad$ $2 \times 12 = 24$ $\qquad$ $3 \times 12 = 36$

$4 \times 12 = 48$ $\qquad$ $5 \times 12 = 60$ $\qquad$ $6 \times 12 = 72$

The first six multiples of 12 are 12, 24, 36, 48, 60, and 72.

Common multiples are multiples shared by two different numbers. The smallest common nonzero multiple shared by two numbers is the **least common multiple** (LCM) of the numbers.

B. Find the least common multiple of 8 and 20.

List the multiples.

8: 8, 16, 24, 32, (40), 48

20: 20, (40), 60, 80, 100, 120

The least common multiple of 8 and 20 is 40.

You can also use prime factorization.

$8 = 2 \times 2 \times 2$ Write the prime factorization of each number.

$20 = 2 \times 2 \times 5$ Multiply the factors using the common factors only once.

LCM = $2 \times 2 \times 2 \times 5$

The LCM is $2 \times 2 \times 2 \times 5$, or 40.

Exercises

Find the least common multiple of each set of numbers.

1. 7 and 10
2. 2 and 9
3. 8 and 14
4. 2 and 7
5. 5 and 14
6. 12 and 16
7. 20 and 50
8. 12 and 15
9. 5 and 15
10. 2, 4, and 9
11. 8, 28, and 30
12. 4, 8, and 15

8 Plotting Points on a Coordinate Plane

Numbers are graphed on the number line. **Ordered pairs** of numbers, such as (3, -4), are graphed on the **coordinate plane**. The coordinate plane has two perpendicular number lines. Each number line is called an **axis**. The axes (plural of *axis*) intersect at the **origin** (0, 0).

Every point in the coordinate plane has two coordinates. The first coordinate is the *x*-**coordinate**, and the second coordinate is the *y*-**coordinate**.

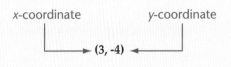

Examples

A. Write the ordered pair that names point P.

1. Start at the origin.
2. Move left on the *x*-axis to find the *x*-coordinate of point P, which is -2.
3. Move up along the *y*-axis to find the *y*-coordinate of point P, which is 3.

The ordered pair for point P is (-2, 3).

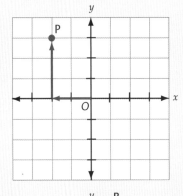

B. Graph and label the ordered pair B(1, 4) on a coordinate plane.

1. Start at the origin.
2. The *x*-coordinate is 1, so move right 1 unit.
3. The *y*-coordinate is 4, so move up 4 units.
4. Draw and label a dot.

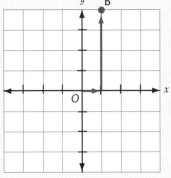

Exercises

Name the ordered pair for the coordinates of each point on the coordinate plane to the right.

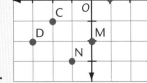

1. B
2. J
3. L
4. C
5. M
6. A
7. T
8. G
9. P
10. R
11. D
12. N

Use graph paper to graph each point on the same coordinate plane.

13. A(2, -4)
14. M(2, 0)
15. K(-4, 4)
16. D(1, -2)
17. E(-1, 2)
18. R(2, 4)
19. L(0, 4)
20. P(-3, -4)

8 Plotting Points on a Coordinate Plane   **xxiii**

9 Converting Units within the Metric System

All units of length in the **metric system** are defined by the **meter** (m). The figure below shows the relationships between some common metric units.

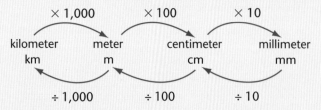

To convert from larger units to smaller units, multiply.

To convert from smaller units to larger units, divide.

Examples

A. Change 150 millimeters to meters.

You are changing from a smaller unit to a larger unit, so divide.

150 mm = _____ m

150 ÷ 1,000 = 0.15   To convert from millimeters to meters, divide by 1,000.

150 mm = 0.15 m

The basic unit of **capacity** in the metric system is the liter (L). A liter (L) and milliliter (mL) are related like the meter and millimeter.

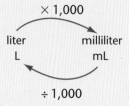

The **mass** of an object is the amount of matter in an object. The basic unit of mass is the gram. The most common metric mass units used are kilogram (kg), gram (g), and milligram (mg). The units are related like the units of length and capacity.

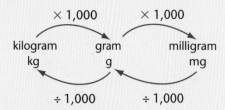

B. Change 644 grams to milligrams.

You are changing from a larger unit to a smaller unit, so multiply.

644 g = _____ mg

644 • 1,000 = 644,000

644 g = 644,000 mg

To convert from grams to milligrams, multiply by 1,000.

C. Change 3.8 liters to milliliters.

You are changing from a larger unit to a smaller unit, so multiply.

3.8 L = _____ mL

3.8 • 1,000 = 3,800

3.8 L = 3,800 mL

To convert from liters to milliliters, multiply by 1,000.

Exercises

State which metric unit you would use to measure each item.

1. a book
2. an eraser
3. a living room floor
4. a paper clip
5. a bucket
6. a bathtub
7. an apple
8. a glass of ice tea

Complete.

9. 4,390 mm = _____ m
10. 63 mL = _____ L
11. 240 mg = _____ kg
12. 915 kg = _____ g
13. 5.35 L = _____ mL
14. 45 g = _____ kg
15. 8 mm = _____ cm
16. 138.5 g = _____ kg
17. 0.85 L = _____ mL
18. 48 mg = _____ g
19. 680 mL = _____ L
20. 6.8 m = _____ cm
21. 4.2 cm = _____ mm
22. 4,250 g = _____ kg

10 Converting Measurements within the Customary System

The most common units of length in the customary system are inch, foot, yard, and mile. The table below shows some common length conversions you will use.

Customary Units of Length
1 foot (ft) = 12 inches (in.)
1 yard (yd) = 3 feet or 36 inches
1 mile (mi) = 5,280 feet or 1,760 yards

To convert from larger units to smaller units, multiply.

To convert from smaller units to larger units, divide.

Examples

A. Change 48 inches to feet.

48 in. = _____ ft

48 ÷ 12 = 4   To convert from inches to feet, divide by 12.

48 in. = 4 ft

B. Change 2.5 miles to feet.

2.5 mi = _____ ft

2.5 • 5,280 = 13,200   To convert from miles to feet, multiply by 5,280.

2.5 mi = 13,200 ft

You can also convert units of capacity and weight. The tables below show the common conversions for capacity and weight.

Customary Units of Capacity
1 cup (c) = 8 fluid ounces (fl oz)
1 pint (pt) = 2 cups
1 quart (qt) = 2 pints
1 gallon (gal) = 4 quarts

Customary Units of Weight
1 pound (lb) = 16 ounces (oz)
1 ton (T) = 2,000 pounds

More Examples

C. Change 12 cups to pints.

12 c = _____ pt

12 ÷ 2 = 6   To convert from cups to pints, divide by 2.

12 c = 6 pt

D. Change 8 pounds to ounces.

8 lb = _____ oz

8 × 16 = 128   To convert from pounds to ounces, multiply by 16.

8 lb = 128 oz

Exercises

Complete.

1. 15 gal = _____ qt
2. 21 yd = _____ in.
3. 8 yd = _____ ft
4. 4 c = _____ fl oz
5. 64 qt = _____ gal
6. 25 mi = _____ ft
7. 24 in. = _____ ft
8. 20 gal = _____ c
9. 152 pt = _____ gal
10. 30 lb = _____ oz
11. 16 lb = _____ oz
12. 6 ft = _____ yd
13. 5 T = _____ lb
14. 104 pt = _____ gal
15. 5 mi = _____ ft
16. 15,840 ft = _____ mi
17. 6,000 lb = _____ T
18. 14,080 yd = _____ mi
19. 6 pt = _____ qt
20. 120 in. = _____ ft

Solve.

21. How many cups are in a quart?
22. How many inches are in a yard?
23. How many ounces are in a pound?
24. How many pints are in a gallon?

Diagnostic Skills Posttest

Replace each ■ with <, >, or = to make a true statement. (Section 1)

1. 2.03 ■ 3.02
2. 61.06 ■ 62.03
3. 2.5 ■ 2.77
4. 0.8 ■ 0.89
5. 0.06 ■ 0.66
6. 1.64 ■ 1.38
7. 10.01 ■ 10.11
8. 1.94 ■ 1.940

Order each set of decimals from least to greatest. (Section 1)

9. 2.2, 3.23, 2.31, 3.32
10. 0.66, 0.61, 0.84, 0.841

Find each sum or difference. (Section 2)

11. 3.5 + 5.3
12. 15.2 + 31.4
13. 61.3 + 21.3
14. 36.2 − 15.1
15. 7.84 − 7.7
16. 4.65 − 3.0
17. 53.6 + 51.4
18. 8.34 − 5.24
19. 43.84 − 28.1
20. 6.12 + 0.60 + 2.29
21. 41.2 − 1.376
22. 0.39 − 0.328

Find each product or quotient. (Section 3)

23. 2.54 • 3.6
24. 42 ÷ 0.7
25. 23.7 • 3.9
26. 82.9 • 0.14
27. 5.76 ÷ 3.6
28. 5.64 ÷ 4.7
29. 24.2 • 0.25
30. 7.24 • 0.36
31. 51.3 • 0.42
32. 57.6 ÷ 2.4
33. 9.49 ÷ 7.3
34. 0.225 ÷ 1.5

Determine whether each number is divisible by 2, 3, 4, 5, 6, 9, or 10. (Section 4)

35. 81
36. 45
37. 223
38. 74
39. 931
40. 632
41. 83
42. 445

Determine whether each number is *prime* or *composite*. Write the prime factorization of each composite number. (Section 5)

43. 21
44. 30
45. 18
46. 56
47. 64
48. 17
49. 39
50. 80

Find the greatest common factor of each set of numbers. (Section 6)

51. 4 and 11
52. 2 and 30
53. 13 and 26
54. 18 and 45
55. 14 and 20
56. 12, 18, and 36

Find the least common multiple of each set of numbers. (Section 7)

57. 6 and 5
58. 8 and 4
59. 10 and 8
60. 7 and 3
61. 6 and 9
62. 3, 4, and 5

Name the ordered pair for the coordinates of each point on the coordinate plane. (Section 8)

63. A
64. B
65. C
66. D
67. E
68. F

On graph paper, graph each point on the same coordinate plane. (Section 8)

69. M(-2, 4)
70. H(3, -2)
71. N(0, 3)
72. T(-4, 0)

Complete. (Section 9)

73. 7,000 m = _____ km
74. 2.6 kg = _____ g
75. 700 g = _____ kg
76. 6,100 mL = _____ L
77. 0.4 L = _____ mL
78. 500 cm = _____ m

Complete. (Section 10)

79. 6 lb = _____ oz
80. 2 pt = _____ c
81. 5 mi = _____ yd
82. 72 in. = _____ yd
83. 6,000 lb = _____ T
84. 12 gal = _____ qt

CHAPTER 1

Integers and Algebra

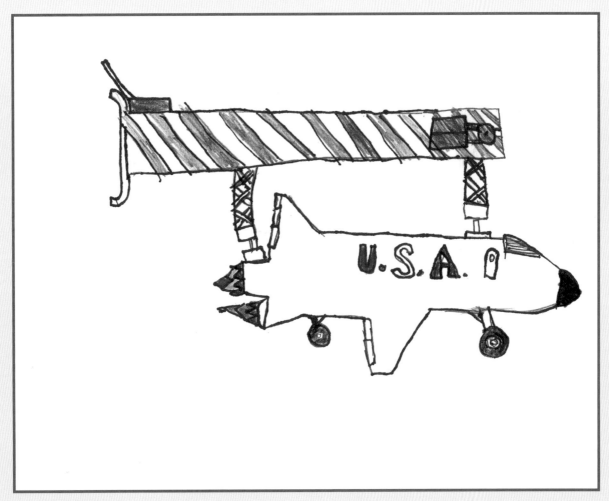

Catherine Beehler
Calvert Day School

1.1 Order of Operations

Objective: to use order of operations

Pat and Phil need 24 pounds of bacon for the Band Booster's Breakfast. They bought twelve 12-ounce packages of bacon and fifteen 16-ounce packages. Who calculated the correct amount?

Pat's calculation

12 • 12 + 15 • 16

144 + 240

384 ounces

384 ÷ 16 = 24 pounds

Phil's calculation

12 • 12 + 15 • 16

144 + 15 • 16

159 • 16

2,544 ounces

2,544 ÷ 16 = 159 pounds

To make certain that an expression, like 12 • 12 + 15 • 16, has only one value, use the following order of operations.

Order of Operations

1. Perform all the operations within parentheses or brackets. Work from the inside out.
2. Evaluate all powers and/or roots.
3. Work all multiplication and/or division from left to right.
4. Work all addition and/or subtraction from left to right.

Pat's calculation is correct.

More Examples

A. 45 − 13 • 4 ÷ 2
   45 − 52 ÷ 2
   45 − 26
   19

B. 44 ÷ [2^2 • (12 − 1)]
   44 ÷ [4 • 11]
   44 ÷ 44
   1

C. (4 + 5) ÷ ($\sqrt{16}$ − 2)
   9 ÷ (4 − 2)
   9 ÷ 2
   4.5

D. $(7 − 3)^2$
   4^2
   16

E. 71 − 17 + $\sqrt{4}$
   54 + 2
   56

Try THESE

For each expression, name the operation that should be completed first. Then evaluate.

1. 18 + 6 − 3

2. 12 + 5 • 6

3. 3 • 2^2 + 1

Exercises

Evaluate.

1. $6 \cdot 3 \div \sqrt{9} - 1$
2. $14 - 3 \cdot 4 \div 6$
3. $36 \div 3^2$
4. $4^2 + 2^3 \cdot \sqrt{25}$
5. $9 \cdot (7 - 4)^2$
6. $(40 \cdot 2) - (6 \cdot 11)$
7. $[(36 - 6) \times 2] + 2^2$
8. $24 \div [18 \div (12 - 3)]$
★9. $4^3 \div (8 + 2^3) \div \sqrt{4}$
★10. $2^4 \cdot [(3^3 - \sqrt{49}) \div \sqrt{64}]$

Insert grouping symbols to make each number sentence true.

11. $7 \cdot 8 - 6 + 3 = 47$
12. $3 + 8 - 2 \cdot 5 = 45$

Compare. Replace each ■ with >, <, or = to make a true statement.

13. $12 \cdot 3 - 5$ ■ $12 \cdot (5 - 3)$
14. $(19 - 15) \div 3 + 1$ ■ $19 - 15 \div 3 + 1$

Problem SOLVING

15. How many ways can you make a total of 50¢ using combinations of pennies, nickels, and dimes?

★16. Benny has a stack of less than 50 pennies. When he divides the number of pennies by 2, by 3, or by 7, there is 1 penny left over. How many pennies does Benny have?

Constructed RESPONSE

17. Show that you can find at least three different values for $3 + 4 \cdot 7 - 1$ if you do not use the rules for the order of operations. Which value is correct? Explain your reasoning.

Mind BUILDER

Patterns

Some numbers can be represented by figures, such as triangles or squares. Study the patterns below.

Triangular Numbers

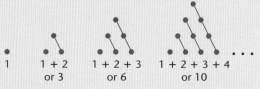

1   1 + 2   1 + 2 + 3   1 + 2 + 3 + 4 . . .
     or 3   or 6   or 10

Square Numbers

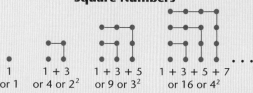

1   1 + 3   1 + 3 + 5   1 + 3 + 5 + 7 . . .
or 1   or 4 or 2^2 or 9 or 3^2 or 16 or 4^2

1. Draw representations, and then list the next three numbers in each pattern.
2. Show, with geometric representations and with numbers, that the sum of two consecutive triangular numbers is equal to a square number.

1.2 Variables

Objective: to evaluate expressions

The value of an expression, such as $b - 7$, may be changed by replacing b with different numbers. In this expression, the letter b is called a **variable**. A variable is used to stand for some number.

Evaluate $b - 7$ if $b = 16$.   Evaluate $b - 7$ if $b = 131.5$.

$b - 7 = 16 - 7$ Replace b with 16. $b - 7 = 131.5 - 7$ Replace b with 131.5.

$\quad\quad = 9$ $\quad\quad = 124.5$

An expression may contain more than one variable. To evaluate the expression, replace each variable with its given value. Then use the order of operations.

More Examples

A. Evaluate $2r + 3t$ if $r = 2.5$ and $t = 6$.

$\quad 2r + 3t = 2 \cdot 2.5 + 3 \cdot 6$ Replace r with 2.5 and t with 6.

$\quad\quad\quad\quad = 5 + 18$ First multiply.

$\quad\quad\quad\quad = 23$ Then add.

$2r$ means "$2 \times r$."
$3t$ means "$3 \times t$."

B. Evaluate $[(s + t) + \sqrt{v}] \cdot 3$ if $s = 21$, $t = 7$, and $v = 4$.

$\quad [(s + t) + \sqrt{v}] \cdot 3 = [(21 + 7) + \sqrt{4}] \cdot 3$ Replace s with 21, t with 7, and v with 4.

$\quad\quad\quad\quad\quad\quad = [28 + \sqrt{4}] \cdot 3$ Work within the parentheses.

$\quad\quad\quad\quad\quad\quad = [28 + 2] \cdot 3$ Evaluate the root.

$\quad\quad\quad\quad\quad\quad = 30 \cdot 3$ Work within the brackets.

$\quad\quad\quad\quad\quad\quad = 90$ Multiply.

Try THESE

Evaluate.

1. $y + 6$, if $y = 5$
2. $5 - x$, if $x = 1$
3. $4a$, if $a = 3$
4. $c \div 2$, if $c = 6$
5. $x + y$, if $x = 2.1$ and $y = 4$
6. $2g - h$, if $g = 3$ and $h = 0.5$

Exercises

Evaluate each expression if $a = 2$, $b = 5$, $n = 0.3$, $x = 4$, and $y = 1.5$.

1. $y + 3$
2. $12 - b$
3. $a - 2$
4. $6x$
5. $9n$
6. $a \div 8$
7. $54 \div y$
8. $6a + 5$
9. $a^2 - 3$
10. $10 - 4n$
11. $6n \div 2$
12. $2 - n^2$
13. $4a + 5b$
14. $\sqrt{x} - y$
15. $y + \sqrt{x}$
16. $4x - y$
17. $a - 5n$
18. $7a - b - 7$
19. $ax + y$
20. $b - x^2 \div 8$
21. $2x + n^2 - b$

Evaluate each expression if $r = 3$, $s = 9$, $t = 0.5$, and $u = 0.3$.

22. $r + (s - u)$
23. $7t \cdot (\sqrt{s} + r)$
24. $2u \div (3t - 1)$
25. $(s - r)^2$

Evaluate each expression if $a = 4$, $b = 2$, and $c = 3$.

★26. $\dfrac{7 + b}{c}$
★27. $\dfrac{13 - a}{b + 7}$
★28. $\dfrac{6c - 2}{a + 2b}$
★29. $\dfrac{(3 + a) \cdot 5}{7b}$

EXAMPLE
$\dfrac{b + 10}{c - 2} = \dfrac{2 + 10}{3 - 2} = \dfrac{12}{1}$ or 12
Evaluate the numerator and denominator separately. Then evaluate the fraction.

Problem SOLVING

30. Jacob Malley sells aluminum siding to homeowners. He earns $975 per month plus $175 per sale. How many sales must he make to earn $2,025 in 1 month?

31. The expression $1.20 + $0.30m can be used to find the cost of a long-distance phone call. The time of the call in minutes is m. Find the cost of a 15-minute call.

★32. A loan of $3,200 is to be repaid in four payments. Each payment after the first is to be $200 greater than the previous payment. What will be the amount of the last payment?

Test PREP

33. $3^2 \cdot 4 + 5$ simplifies to ____.   a. 81   b. 41   c. 17   d. 29
34. $[5 + (8 - 3) \cdot 2] + 1$ simplifies to ____.   a. 16   b. 21   c. 26   d. 8

1.3 Expressions and Equations

Objective: to translate phrases and sentences into algebraic expressions and equations

Juanita makes a cake for a bake sale. The icing contains 2 times as much sugar as the cake.

THINK
If the cake contains 1 cup of sugar, then the icing uses 2 • 1 cups of sugar.

If the icing contains 4 cups of sugar, then the cake uses 4 ÷ 2 cups of sugar.

THINK
If the cake contains c cups of sugar, then the icing uses 2 • c cups of sugar.

If the icing contains w cups of sugar, then the cake uses w ÷ 2 cups of sugar.

You can translate words into **algebraic expressions**. Use variables to represent unnamed numbers.

Words	Expressions
some number increased by 6	$h + 6$
the difference of b and 9	$b - 9$
five multiplied by some number	$5z$
the quotient of 8 and n	$8 \div n$ or $\frac{8}{n}$
fourteen minus a number	$14 - q$

More Examples

An **equation** is a mathematical sentence with an equals sign. You can translate many sentences into equations.

Sentences	Equations
A. Five more than x is seventeen.	$x + 5 = 17$
B. A number decreased by 4 is 25.	$n - 4 = 25$
C. The product of 8 and a number is 60.	$8q = 60$
D. Eighteen divided by d is three.	$18 \div d = 3$
E. Two less than one-fourth of a number equals three.	$\frac{1}{4}h - 2 = 3$
F. Two times the sum of a number and four is fourteen.	$2(b + 4) = 14$

$$2 \times (b+4) = 14$$

Try THESE

Translate each phrase into an algebraic expression.

1. seventeen more than a
2. ten less than p
3. five times k
4. the quotient of t and 29
5. the difference of a number and 16
6. a number increased by 4
7. a number divided by 6
8. the product of 215 and a number

Exercises

Translate each sentence into an equation.

1. The sum of m and 15 is 45.
2. Five decreased by r is two.
3. The quotient of d and 7 is 6.
4. The product of 5 and y is 30.
5. Six less than a number is eight.
6. Two more than a number is six.
7. Seven times a number is twenty-one.
8. A number divided by 4 is 5.
9. Twelve subtracted from n is nine.
10. Five-tenths of x is thirty.
11. Three more than four times y is equal to twenty-seven.
12. Six less than twice g is equal to fourteen.
13. Five times the difference of a number and eighteen is forty-two.
14. The sum of 6 times a number and 3 is 48.

Translate into words.

15. $b + 6$
16. $8 - e$
17. $r \div 7$
18. $18a$
19. $14g = 7$
20. $d - 1 = 42$
21. $100 \div t = 5$
22. $26 + \frac{1}{2}v = 9$

Problem SOLVING

23. The Cubs scored c points. The Bucks scored 2 more points than the Cubs. Write an expression for the number of points scored by the Bucks.
24. There were t questions on a test. Sam missed 3 questions. Write an expression for the number of questions he answered correctly.
★ 25. Write an expression for the number of hours in s seconds.
26. Write an expression for the number of nickels that equals d dollars.

MIXeD REVIEW

Replace each n with a number to make a true sentence.

27. $n + 3 = 3$
28. $1 \bullet 4.5 = n$
29. $248 = n + 0$
30. $(5 + 7) + n = 3 + (5 + 7)$
31. $0.5 \bullet (10 + 3) = (n \bullet 10) + (n \bullet 3)$

1.4 Properties

Objective: to identify and use properties

Janna is multiplying three numbers: 12, 7, and 5. She knows that she can rearrange the order of the numbers to be able to evaluate the numbers more easily. She can use the properties below.

Commutative Property

Addition	Multiplication
The order in which the addends are added does not change the sum.	The order in which the factors are multiplied does not change the product.
$3 + 4 = 4 + 3$	$5 \cdot 2 = 2 \cdot 5$
$a + b = b + a$	$a \cdot b = b \cdot a$

You can also change the grouping of the numbers before adding or multiplying them without changing the sum or product.

Associative Property

Addition	Multiplication
The way in which the addends are grouped does not change the sum.	The way in which the factors are grouped does not change the product.
$(3 + 4) + 5 = 3 + (4 + 5)$	$(5 \cdot 2) \cdot 8 = 5 \cdot (2 \cdot 8)$
$(a + b) + c = a + (b + c)$	$(a \cdot b) \cdot c = a \cdot (b \cdot c)$

When you add 0 to a number or multiply a number by 1, the number does not change.

Identity Property

Addition	Multiplication
When 0 is added to a number, the sum is that number.	When a number is multiplied by 1, the product is that number.
$7 + 0 = 7$	$8 \cdot 1 = 8$
$a + 0 = a$	$a \cdot 1 = a$

Examples

A. Name the property.

$10 + 15 = 15 + 10$ Order is changed.

This is an example of the commutative property of addition.

B. Fill in the missing number to make the number sentence true.

$4.7 = n \cdot 4.7$

$4.7 = 1 \cdot 4.7$ identity property of multiplication

The missing number is 1.

C. Simplify. $14 + (6 + 9)$

$14 + (6 + 9) = (14 + 6) + 9$ associative

$= 20 + 9$ parentheses

$= 29$ Add.

D. Simplify. $68 + 35 + 22$

$68 + 35 + 22 = 68 + 22 + 35$ commutative

$= 90 + 35$ Add.

$= 125$ Add.

Try THESE

Match the equation with the property it illustrates.

1. $\frac{2}{3} \cdot \frac{1}{2} = \frac{1}{2} \cdot \frac{2}{3}$

2. $7 \cdot 1 = 7$

3. $(2 \cdot 3) \cdot 4 = 2 \cdot (3 \cdot 4)$

4. $0 + \frac{3}{4} = \frac{3}{4}$

a. identity property of multiplication

b. associative property of multiplication

c. commutative property of multiplication

d. identity property of addition

Exercises

Name the property shown.

1. $2\frac{1}{2} + 0 = 2\frac{1}{2}$

2. $\left(\frac{1}{2} \cdot \frac{1}{3}\right) \cdot \frac{2}{5} = \frac{1}{2} \cdot \left(\frac{1}{3} \cdot \frac{2}{5}\right)$

3. $35 \cdot 1 = 35$

4. $10 + 15.5 = 15.5 + 10$

Replace each n with a number to make a true sentence.

5. $3 + n = 3$

6. $5 \cdot 4 = n \cdot 5$

7. $(1 + 3) + 5 = n + (3 + 5)$

8. $0 + \frac{2}{5} = \frac{2}{5} + n$

Problem SOLVING

9. You want to paint the walls in your family room. The walls are 9 feet tall, and they are 12 feet, 10.5 feet, 11 feet, and 15 feet wide. (*Remember:* area = length × width)

   a. Write two expressions to find the area of the walls.

   b. Then find the total area.

Constructed RESPONSE

10. Is division commutative? Explain your reasoning using your knowledge of the commutative property, and give an example.

11. Is the expression "$14 + 18 - 25 + 31$" equal to "$18 + 31 - 25 + 14$"? Explain your reasoning.

Problem Solving

Guess Again

Aaron's Auto World is promoting their grand opening by giving away a new car. Each customer is allowed to guess how many car keys are in a large jar. Gary and his five brothers made the following guesses.

Gary	594
Larry	600
Barry	589
Harry	591
Perry	597
Terry	593

One of the brothers won the car by guessing the correct number. Two brothers missed by 3. One brother missed by 6. Another brother missed by twice as much as Terry missed. Which brother won the new car?

Extension

What are several methods you could use to make an educated guess on the number of car keys in the jar?

Cumulative Review

Round to the underlined place-value position.

1. <u>7</u>6
2. 0.4<u>5</u>
3. <u>1</u>.83
4. 0.0<u>6</u>2
5. <u>2</u>9.91
6. 8<u>5</u>,243
7. 1<u>5</u>,396,000
8. 7.7<u>8</u>17

Compute.

9. 545 + 79
10. 8,606 + 1,295
11. $6.46 + 5.59
12. 16.5 + 8.42

13. 1,862 − 371
14. $86,010 − 24,704
15. 45.3 − 9.6
16. 718 × 4

17. 416 × 209
18. $72 × 1.1
19. 0.43 × 0.06

20. 8)‾70
21. 28)‾1.008
22. 9)‾7.38
23. 24.6 + 7.45
24. 461 + 84 + 72.05
25. 114,018 − 11,699
26. 673.01 − 8.076
27. $81 − $4.83
28. $56 • 30
29. 31.5 • 6
30. 0.053 • 0.5
31. 684 ÷ 1.9
32. 0.135 ÷ 0.05

Solve. Use the mileage chart for problems 33–35.

33. Which trip is shorter, a trip from Louisville to Miami through Atlanta or a trip from Louisville to Miami through Memphis?

34. How many miles longer is a trip from Dallas to Miami than a trip from Louisville to Atlanta?

35. A trip from Memphis to New Orleans takes 13 gallons of fuel. Find the miles per gallon.

36. Elizabeth buys 5 pairs of socks for $2.98 each. The sales tax is $0.82. What is her change from two ten-dollar bills?

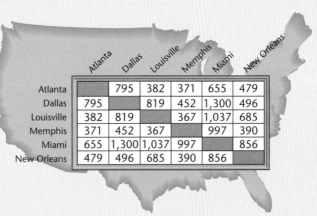

	Atlanta	Dallas	Louisville	Memphis	Miami	New Orleans
Atlanta		795	382	371	655	479
Dallas	795		819	452	1,300	496
Louisville	382	819		367	1,037	685
Memphis	371	452	367		997	390
Miami	655	1,300	1,037	997		856
New Orleans	479	496	685	390	856	

1.5 Integers and Absolute Value

Objective: to find the opposite and absolute value of integers

The average low temperature in International Falls, Minnesota, in January is 8°F below zero. You can express this temperature using the negative number -8.

Numbers, such as 8 and -8, are **opposites** because they are the same distance from 0. The set of whole numbers and their opposites are **integers**.

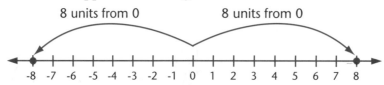

A part of the set of integers is listed below. **Negative integers** are to the left of 0, and **positive integers** are to the right of 0.

$$\ldots, -5, -4, -3, -2, -1, 0, 1, 2, 3, 4, 5, \ldots$$

Examples

A. The opposite of 2 is -2.   **B.** The opposite of -5 is 5.

On the number line above, you can see that 8 and -8 are different numbers. However, each is 8 units from 0. The number of units a number is from 0 on a number line is called its **absolute value**. The absolute value symbol is $|\ |$.

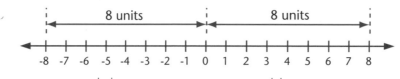

$|-8| = 8$ $|8| = 8$

More Examples

C. $|-4| = 4$ **D.** $|12| = 12$ **E.** $|-2| = 2$

F. Evaluate $5 + |h|$ when $h = -6$.

$5 + |h| = 5 + |-6|$ Replace h with -6.

$\quad\quad\ \ = 5 + 6$ $|-6| = 6$

$\quad\quad\ \ = 11$ Simplify.

You can use the number line above to compare and order integers. Integers increase from left to right on a number line.

G. Compare. -3 ■ 2

Draw a number line, and plot the integers.

-3 -2 -1 0 1 2 3

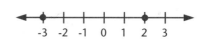

-3 is to the left of 2, so -3 < 2.

Try THESE

Write an integer to describe each of the following.

1. 5°F above zero
2. $500.00 loss
3. 14°F below zero
4. 280 feet above sea level
5. down 8 floors
6. up 6 floors

Exercises

State the opposite of each integer.

1. 4
2. -9
3. 42
4. 0
5. 1
6. -3
7. -10
★8. -(-6)

Evaluate each expression.

9. $|6|$
10. $|-9|$
11. $|-5| + |6|$
12. $|0|$
13. $|12 - 4|$
14. $|-9| - |6|$
15. $|-4| + |-7|$
16. $|3| + |-8|$

Evaluate each expression if $a = 4$, $b = -6$, and $c = -2$.

17. $|c| + 5$
18. $a + |b|$
19. $|b| + |c|$
20. $5|b|$

Replace each ■ with <, >, or = to make a true statement.

21. 3 ■ 5
22. -5 ■ 0
23. -7 ■ 3
24. 0 ■ -2
25. 6 ■ -2
26. -5 ■ -8
27. -6 ■ $|-8|$
28. $|-18|$ ■ 0

Order the integers in each list from least to greatest.

29. -3, 5, -1
30. 13, -12, -6, 5
31. -8, 8, 6, $|-7|$, 5
32. 3, $|2|$, -1, 0, $|-4|$

Problem SOLVING

Use the lake elevations table to the right to solve each problem.

33. Which lake is at a higher elevation, Lake Smith or Bush Lake?
34. List the lakes in order of elevation from least to greatest.

Lake	Elevation (ft)
Red Lake	15
Lake Atley	0
Lake Smith	-22
Bush Lake	-10

Test PREP

35. Which set of numbers is in order from least to greatest?
    a. -1, 5, -3
    b. 2, -3, 4
    c. -3, 2, 3
    d. 0, -2, 5

36. Which expression has the greatest value?
    a. $|5 - 4|$
    b. $|-6|$
    c. $|3 + 2|$
    d. $2|-2|$

1.6 Adding and Subtracting Integers

Objective: to add and subtract integers

The temperature at 7:00 A.M. is -4°C. The temperature rises 7°C by noon. What is the temperature at noon?

-4 + 7 = ?

Addition can be shown on a number line. Move right when the sign is positive. Move left when the sign is negative.

-4 + 7 = 3

The temperature at noon is 3°C.

Do you think the temperature at noon is greater or less than 0?

Start at 0.
Move 4 units to the left.
From there move 7 units to the right.

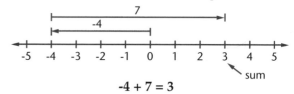

-4 + 7 = 3

More Examples

A. -5 + 4 = -1   The difference between 5 and 4 is 1.

B. 5 + -4 = 1

In each example, the sum has the same sign as the number with the greater absolute value. These examples suggest the following rule.

 Rule To add integers with different signs, find the absolute value of each. Subtract the lesser absolute value from the greater absolute value. Give the result the sign of the number with the greater absolute value.

C. Find -3 + -2.

-3 + -2 = -5

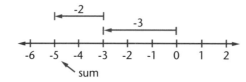

Start at 0.
Move 3 units to the left.
From there move 2 more units to the left.

 Rule To add integers with the same sign, add their absolute values. The sum has the same sign as the numbers.

Any subtraction sentence can be changed to a corresponding addition sentence.

Subtraction   **Addition**

⎯⎯ opposites ⎯⎯

D. 14 − 8 = 6   14 + -8 = 6

E. 5.5 − 2 = 3.5   5.5 + -2 = 3.5

⎯⎯ same results ⎯⎯

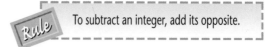 **Rule** To subtract an integer, add its opposite.

Try THESE

Write each subtraction expression as an addition expression. Then find each difference.

1. $6 - 10$
2. $7 - -6$
3. $-8 - -8$
4. $-9 - 5$
5. $18 - -18$
6. $25 - 25$
7. $-2.6 - 1.9$
8. $0.5 - 0.8$

Exercises

Add or subtract.

1. $6 + 8$
2. $-6 + 8$
3. $6 + -8$
4. $-6 + -8$
5. $4 - 11$
6. $-9 - 8$
7. $15 + -6$
8. $-2 - -4$
9. $8 - -6$
10. $3 + -15$
11. $-6.4 - -5.8$
12. $-8 - -14$
13. $-6 + -6$
14. $-34.3 + -24.1$
15. $7 - 25$
16. $1.8 + -3.1$
17. $18 - 7$
18. $-36 + 0$
19. $4 - |-5|$
20. $|-10| - |18|$

Evaluate each expression if $x = -5$, $y = 3$, and $z = -4$.

21. $3 + x$
22. $z + 9$
23. $y - 7$
24. $x - y$
25. $|x| + -6$
26. $|x - y|$
27. $|z| + y$
28. $x + y + z$

Add.

29. $-3 + 7 + 3$
30. $2 + -8 + -12$
31. $15 + -12 + -18$
32. $-25 + 32 + -15$

Write an addition sentence for each problem. Then find the sum.

33. The temperature drops 7 degrees and then rises 11 degrees.
34. Ricardo loses 15 pounds, gains 6 pounds, and then loses 10 pounds.

Problem SOLVING

35. Mount Everest is the highest point in the world at 8,848 feet. The Dead Sea is the lowest point on land at -400 feet. What is the difference in their elevations?

36. Does $|a + b|$ equal $|a| + |b|$ when a and b are positive integers? How about when a and b are negative numbers? How about when a is positive and b is negative? Explain your reasoning using your knowledge of absolute value and integers.

Mixed REVIEW

Simplify.

37. $3 + 5 \cdot 8$
38. $4 + (8 \cdot 3 - 1)$
39. $5^2 - (2 + 1)$
40. $[(12 - 8) + 5] + 6$
41. $4^2 + 4 \div 4$
42. $18 \div 3^2 + 5$

1.7 Multiplying and Dividing Integers

Objective: to multiply and divide integers

Wave erosion may cause a coastline to recede. Suppose a certain beach loses 2 centimeters each year, or -2. The beach loses 3 • -2, or -6, centimeters in 3 years.

You can think about multiplication as repeated addition. So, -2 is used as an addend 3 times.

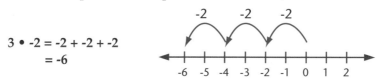

3 • -2 = -2 + -2 + -2
= -6

The commutative property states that the order of the factors does not change the product.

3 • -2 = -6 and -2 • 3 = -6

> The product of two integers with different signs is negative.

The product of two positive integers is positive. For example, 3 • 4 = 12. What is the sign of the product of two negative integers?

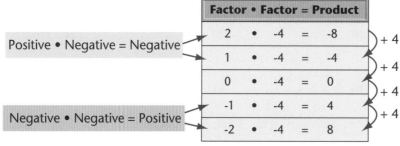

This example suggests the following rule.

> The product of two integers with the same sign is positive.

Examples

A. Find -5 • -9.

-5 • -9 = 45   The factors have the same sign. The product is positive.

B. Find 6 • -3.

6 • -3 = -18   The factors have different signs. The product is negative.

In mathematics, you know that division undoes multiplication. Look at the following multiplication sentences and their related division sentences.

8 • 4 = 32   so   32 ÷ 4 = 8 (same sign)
7 • -10 = -70   so   -70 ÷ -10 = 7 (same sign)

> In each division sentence, the signs of the dividend and divisor are the *same*. Each quotient is *positive*.

-8 • 10 = -80   so   -80 ÷ 10 = -8 (different signs)
-7 • -9 = 63   so   63 ÷ -9 = -7 (different signs)

> In each division sentence, the signs of the dividend and divisor are *different*. Each quotient is *negative*.

> The quotient of two integers with different signs is negative. The quotient of two integers with the same sign is positive.

More Examples

C. Find -50 ÷ 10.

-50 ÷ 10 = -5   The signs are different. The quotient is negative.

D. Find -48 ÷ -6.

-48 ÷ -6 = 8   The signs are the same. The quotient is positive.

Try THESE

State the sign of each product or quotient. Then multiply or divide.

1. -2 • 5
2. 24 ÷ -8
3. 20 ÷ 2
4. 6 • -3
5. -3 • -7
6. -40 ÷ 4
7. -45 ÷ -9
8. -22 • -1

Exercises

Multiply.

1. -9 • 6
2. -5 • -7
3. 16 • -4
4. 26 • -11
5. -12 • 50
6. -9 • -3
7. -3 • -12
8. 5 • 27
9. -3 • 31
10. -14 • -5
11. 13 • 17
12. -24 • -4
13. -2 • 5 • 4
14. (-6)²
15. 6 • -4 • -3
16. 8 • -2 • 4 • 5

Divide.

17. -54 ÷ -9
18. 20 ÷ -5
19. 32 ÷ -2
20. -96 ÷ 12
21. -48 ÷ -3
22. 80 ÷ -6
23. -91 ÷ 13
24. -343 ÷ -7
25. -270 ÷ 18
26. 72 ÷ -8
27. -16 ÷ -1
28. 72 ÷ -6

Replace each ■ with <, >, or = to make a true statement.

29. 8 − 54 ÷ -6 ■ 8 − 9
30. -6 ÷ 3 • -2 ■ -6 • 2
31. 24 − 3 + 9 ■ -3 • -8
32. (-6 − 12) ÷ 32 ■ 6 + -2 • 1

Use integer rules and other math facts to answer each question.

33. What integer and -7 have the product -35?
34. What integer and -6 have the quotient 18?

Problem SOLVING

35. If one factor is positive and the product is negative, what is the sign of the other factor?
36. If an odd number of negative factors are multiplied, is the product positive or negative? Explain and give examples.

Constructed RESPONSE

37. If a number is negative, is the square of the number negative? Explain your reasoning using your knowledge of integers.

Mid-Chapter REVIEW

Evaluate.

1. 4 + 5 • 2 − 3
2. (4 + 8) • 2 ÷ 6
3. $(5^2 − 3) ÷ 2$
4. 3a − 12, if a = 5
5. (4a − 4) ÷ 5, if a = 6
6. $a^2 + b^2$, if a = 5 and b = 4

Replace each n with a number to make a true sentence.

7. 5 • n = 7 • 5
8. 2 + (9 + 3) = (2 + n) + 3
9. 6 • n = 6

Compute.

10. -3 • 8
11. 5 − 9
12. -7 + -12
13. -32 ÷ -8

Problem Solving

Which One is Different?

In each row, one design is different from the others. Write a sentence that explains which one is different and why. (There may be more than one correct answer.)

Extension

Draw a fifth figure for each row having at least one common and one different characteristic of the other four.

1.8 Inverse and Distributive Properties

Objective: to identify and use the inverse and distributive properties

Two properties that are useful when solving equations are the inverse property and the distributive property.

When you add inverses, the sum is 0, and when you multiply **reciprocals**, the product is 1.

Inverse Property

Addition	Multiplication
When a number is added to its opposite, the sum is 0.	When a number is multiplied by its reciprocal, the product is 1.
$3 + -3 = 0$	$2 \cdot \frac{1}{2} = 1$
$a + -a = 0$	$a \cdot \frac{1}{a} = 1$

Distributive Property

The distributive property is another special property. It states that the product of a number and a sum is equal to the sum of the products. It can be applied to an expression involving a sum or difference of two or more numbers or variables.

$$5(3 + 6) = 5 \cdot 3 + 5 \cdot 6 \quad \text{or} \quad a(b + c) = ab + ac$$

Examples

A. Find the product.

$9(x + 5) = 9 \cdot x + 9 \cdot 5$ Use the distributive property.

$ = 9x + 45$ Simplify.

B. Find the product.

$8(2x - 3) = 8 \cdot 2x - 8 \cdot 3$

$ = 16x - 24$

C. You can also use the distributive property to multiply mentally.

$5(98) = 5(90 + 8)$ Write 98 as 90 + 8.

$5(90 + 8) = 5(90) + 5(8)$ Use the distributive property.

$ = 450 + 40$ Multiply.

$ = 490$ Add.

Try THESE

Match the equation with the property it illustrates.

1. 8 + -8 = 0
2. 7(b + 2) = 7 • b + 7 • 2
3. $9 \cdot \frac{1}{9} = 1$

a. inverse property of multiplication
b. inverse property of addition
c. distributive property

Exercises

Replace each *n* with a number to make a true sentence.

1. 4 + -n = 0
2. $n \cdot \frac{1}{5} = 1$
3. $n \cdot \frac{1}{3} = 1$
4. n(4 + 6) = (0.5 • 4) + (0.5 • 6)

Evaluate each expression mentally.

5. 31 + 18 + 9
6. 4(108)
7. $20(\frac{1}{2} \cdot 35)$
8. (18 + 26) + 14
9. 6(23)
10. 5 • 255 • 20

Use the distributive property to find each product.

11. 5(a + 6)
12. 3(a – 7)
13. 10(3g + 9)
14. 3(8 + m)
15. 8(d – 9)
16. 5(6x + 8)

Problem SOLVING

17. Use the numbers 4, 5, and 6 to write an equation that equals 54.

Constructed RESPONSE

18. Is there an inverse property of subtraction? Explain.

Mind BUILDER

Riddle

This is the story of a race between a turtle and a rabbit. They had the following conversation.

Rabbit I can run 10 times as fast as you. I'll give you a 1 km head start and still win.

Turtle But, while you run the 1 km, I'll run another 0.1 km. Then, while you run the 0.1 km, I'll run another 0.01 km, and so on. You will never catch me!

Did the rabbit ever catch the turtle? If so, how far did the turtle run before the rabbit caught up?

1.9 Solving One-Step Equations

Objective: to solve one-step equations

Anna Gaultney uses a scale to weigh vegetables at her market. Suppose the scale is balanced and she adds weight to one pan. Does the scale still balance? What happens if she adds the same weight to each pan? Does the scale still balance?

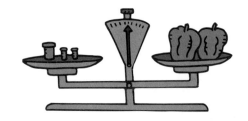

Equations are like scales in balance. If you add or subtract the same number on each side of an equation, the result is an equation that has the same **solution**. It is called an **equivalent equation**.

Subtraction Property of Equality
Subtracting the same number from each side of an equation results in an equivalent equation.

Addition Property of Equality
Adding the same number to each side of an equation results in an equivalent equation.

Division Property of Equality
Dividing each side of an equation by the same nonzero number results in an equivalent equation.

Multiplication Property of Equality
Multiplying each side of an equation by the same number results in an equivalent equation.

Examples

A. $x + \text{-}5 = 3$

$x + \text{-}5 + 5 = 3 + 5$ Add the opposite of -5 to each side.

$x = 8$

Check: $x + \text{-}5 \stackrel{?}{=} 3$ Replace x with 8.

$8 + \text{-}5 \stackrel{?}{=} 3$

$3 = 3$ ✓

The solution is 8.

B. $n - 2.5 = \text{-}4.1$

$n - 2.5 + 2.5 = \text{-}4.1 + 2.5$ Add 2.5 to each side.

$n = \text{-}1.6$

Check: $n - 2.5 \stackrel{?}{=} \text{-}4.1$ Replace n with -1.6.

$\text{-}1.6 - 2.5 \stackrel{?}{=} \text{-}4.1$

$\text{-}4.1 = \text{-}4.1$ ✓

The solution is -1.6.

C. $\text{-}4b = 36$

$\dfrac{\text{-}4b}{\text{-}4} = \dfrac{36}{\text{-}4}$ Divide each side by -4.

$b = \text{-}9$

The solution is -9.

D. $\dfrac{y}{\text{-}12} = \text{-}5$

$\text{-}12 \cdot \dfrac{y}{\text{-}12} = \text{-}12 \cdot \text{-}5$ Multiply each side by -12.

$y = 60$

The solution is 60.

Try THESE

Name the operation and number you would use to solve each equation.

1. $x + 4 = 12$
2. $\text{-}6 = \dfrac{b}{5}$
3. $9 + x = \text{-}8$
4. $4t = 52$
5. $\text{-}6n = 48$
6. $15 = x - 3$
7. $x + \text{-}9 = 12$
8. $6 = \dfrac{b}{\text{-}3}$

Exercises

Solve each equation. Check your solution.

1. $7 + g = -12$
2. $9 = m + -2$
3. $24 = 10 + h$
4. $c + 7 = -16$
5. $h - 7 = 24$
6. $n - -4 = -7$
7. $y + 12 = -3$
8. $-4.2 = 2.3 + h$
9. $5 = v - 9$
10. $7.6 = q - 5.1$
11. $k - 1.6 = -4.3$
12. $-3.1 = x - 1.7$
13. $y + -2 = 5$
14. $b - 7 = -3$
15. $y - 8 = -4$

Translate each sentence into an equation. Then solve the equation.

16. r decreased by 3 is 49.
17. Six is twelve less than m.
18. The sum of -15 and y is 41.
19. Some number increased by -3 is 25.

Solve each equation. Check your solution.

20. $9k = -72$
21. $32 = -8a$
22. $\dfrac{y}{-7} = -8$
23. $21 = \dfrac{h}{6}$
24. $-12d = 48$
25. $\dfrac{h}{-9} = 7$
26. $-0.4c = 9.68$
27. $-144 = -24x$
28. $\dfrac{x}{2.25} = -9$
29. $-4y = -96$
30. $\dfrac{g}{-5} = -20$
31. $-1.2d = -54$

Translate each sentence into an equation. Then solve the equation.

32. Six multiplied by t is negative thirty-six.
33. The quotient of y and -3 is -18.
34. A number divided by -8 is 3.
35. The product of x and -12 is -60.

Problem SOLVING

36. A helicopter descends 160 meters to observe traffic. It levels off at 225 meters. What was the original altitude?

★37. The sum of two numbers is 368. The difference of the same numbers is 86. What are the numbers?

38. Barry and Larry have a combined weight of 288 pounds. Larry weighs 12 pounds more than Barry. How much does Barry weigh?

39. Geoffrey withdraws $25 a week from his bank account. How many weeks will it take him to withdraw $350?

Test PREP

40. How is the sentence "Negative four multiplied by x is one hundred twenty" written as an equation?
    a. $x = -4 \cdot 120$
    b. $-4 + x = 120$
    c. $120 = -4x$
    d. $120x = -4$

41. How is the sentence "Three added to n is negative five" written as an equation?
    a. $3n = -5$
    b. $n + 3 = -5$
    c. $n - 3 = -5$
    d. $3 + -5 = n$

1.10 Solving Two-Step Equations

Objective: to solve two-step equations

In science class, you may convert temperatures between the Celsius and Fahrenheit scales. You can use the **formula** $F = 1.8C + 32$. The weather forecaster states that it will be 77°F today. What is the temperature in degrees Celsius?

You can use the equation $77 = 1.8C + 32$.

This is a two-step equation. A two-step equation contains two operations. To solve it you must complete the inverse operations in the correct order.

To solve a two-step equation, complete the following steps:
    1. Undo the addition or subtraction.
    2. Undo the multiplication or division.

Step 1	Step 2
Undo the addition.	Undo the multiplication.
$77 = 1.8C + 32$	$45 = 1.8C$
$77 - 32 = 1.8C + 32 - 32$	$\dfrac{45}{1.8} = \dfrac{1.8C}{1.8}$
$45 = 1.8C$	$25 = C$

Subtract 32 from both sides.

Divide each side by 1.8.

Check:
$77 = 1.8 \cdot 25 + 32$
$77 = 45 + 32$
$77 = 77$

The temperature in Celsius is 25°.

Examples

A. $5 = \dfrac{m}{3} - 12$

    $5 + 12 = \dfrac{m}{3} - 12 + 12$ Add 12 to both sides.

    $17 = \dfrac{m}{3}$

    $3 \cdot 17 = \dfrac{m}{3} \cdot 3$ Multiply both sides by 3.

    $51 = m$

B. $4x + 6 = -10$

    $4x + 6 - 6 = -10 - 6$ Subtract 6 from both sides.

    $4x = -16$

    $\dfrac{4x}{4} = \dfrac{-16}{4}$ Divide each side by 4.

    $x = -4$

Try THESE

State what to work *first* and what to work *second* to solve each equation. Then solve each equation.

1. $2r - 7 = 1$
2. $\frac{y}{5} + 8 = 7$
3. $4 - 2b = -8$

Exercises

Solve each equation. Check your solution.

1. $2y - 1 = 9$
2. $2y + 1 = -31$
3. $2m + 5 = -29$
4. $-32 = -2 + 5y$
5. $-11 = 3d - 2$
6. $0 = 11y + 33$
7. $-8 + 6m = -50$
8. $13 = -7 + 2y$
9. $1 - 3x = 7$
10. $10 - 2x = -2$
11. $-4y + 3 = 19$
12. $-8k - 21 = 75$
13. $\frac{x}{4} + 6 = 10$
14. $5 + \frac{t}{2} = 3$
15. $6 + \frac{z}{5} = 0$
★16. $\frac{1}{4}t - 2 = 1$
★17. $\frac{2}{3}y + 11 = 33$
★18. $-6 = 4 + \frac{5}{2}y$

Problem SOLVING

19. Logan and Luke are solving the same problem. Who is correct? Explain.

Logan
$3x + 6 = 21$
$ - 6 = -6$
$\frac{3x}{3} = \frac{15}{3}$
$x = 5$

Luke
$\frac{3x + 6}{3} = \frac{21}{3}$
$x + 6 = 7$
$ - 6 = -6$
$x = 1$

Mind BUILDER

Find the Equation

First Number	-2	-1	0	1	2	3	4	10	-10
Second Number	-7	-5	-3	-1	1	■	■	■	■

1. Find the missing values in the table above.
2. How are the second numbers related to the first numbers?
3. If y is the second number and x is the first number, write an equation that represents the second number.

1.11 Problem-Solving Strategy: Write an Equation

Objective: to solve word problems by writing an equation

Maria Hall freezes forty-eight quarts of peaches and cherries. Fourteen of the quarts are peaches. If Mrs. Hall keeps half of the cherries and gives the other half to a neighbor, how many quarts of cherries does she keep?

You can use an equation to solve this problem.

You need to find how many quarts of cherries Mrs. Hall keeps. There are 48 quarts of peaches and cherries. You know there are 14 quarts of peaches. Mrs. Hall keeps half of the cherries.

Let c represent the number of quarts of cherries she keeps. Therefore, $2c$ represents the total number of quarts of cherries. Translate the words into an equation using the variable.

quarts of peaches	plus	quarts of cherries	equals	total quarts
14	+	$2c$	=	48

$$14 - 14 + 2c = 48 - 14 \quad \text{First subtract 14.}$$
$$2c = 34$$
$$\frac{2c}{2} = \frac{34}{2} \quad \text{Then divide by 2.}$$
$$c = 17$$

Mrs. Hall keeps 17 quarts of cherries.

If Mrs. Hall keeps 17 quarts of cherries, what is the total number of quarts of peaches *and* cherries she freezes?

$$14 + 2c = 48$$
$$14 + 2 \cdot 17 = 48 \quad \text{Replace } c \text{ with 17.}$$
$$14 + 34 = 48$$
$$48 = 48 \checkmark$$

The answer is correct.

Try THESE

Translate each problem into an equation. Then solve the equation.

1. Adam Graber has 4 hamsters and some fish. The number of fish divided by 5 is equal to the number of hamsters. How many fish does Adam have?

2. You want to make 3 pineapple cakes for a party. The cakes contain 3 sticks of butter in all. Altogether the cakes and the icing contain $7\frac{1}{2}$ sticks of butter. How much butter is needed for the icing of just 1 cake?

Solve

1. Akiko has 3 fewer rows of tomatoes than rows of beans. She has 5 rows of tomatoes. How many rows of beans does Akiko have?

2. Roy's recipe for punch makes 20 more servings than half of Dixie's recipe. Roy's recipe makes 50 servings. How many servings does Dixie's recipe make?

3. John weighs 66 kilograms. This is 4 kilograms more than twice Atepa's weight. What is Atepa's weight?

4. There are 96 students signed up for the soccer league. They have 16 sponsors. How many teams of 8 players each can be formed?

5. While at the grocery store, you bought 6 pounds of potatoes and a gallon of milk. If the gallon of milk cost $2 and you spent $5, how much did each pound of potatoes cost?

6. Bill runs four times a week. He runs the same distance on Monday, Wednesday, and Thursday. He runs 8 miles on Saturday. If he runs a total of 23 miles a week, how many miles does he run on Wednesday?

7. Write a word problem that can be modeled by the equation $3n - 6 = 150$.

8. You have a dog walking business and earn $8 each time you walk a dog. How many dogs do you have to walk to earn $102 to buy a bike if you have already saved $30?

9. A taxi charges $2 per trip plus $0.50 per mile. If a patron was charged $6.50, how many miles did he travel?

1.11 Problem-Solving Strategy: Write an Equation

Chapter 1 Review

Language and Concepts

Choose a term from the list at the right to correctly complete each sentence.

commutative
opposite
inverse
identity
algebraic expression
integers
variable
associative

1. 62 is the _____ of -62.
2. A symbol used to stand for a number is a(n) _____.
3. 3 • 1 = 3 is an example of the _____ property of multiplication.
4. A(n) _____ is a phrase that contains numbers, variables, and operation symbols.
5. 5 + 7 = 7 + 5 is an example of the _____ property of addition.
6. The set of whole numbers and their opposites are _____.

Skills and Problem Solving

For each expression name the operation that should be completed first. Then evaluate. Section 1.1

7. $18 - 3 \cdot 4$
8. $(5 + 2) \cdot 8 - 28$
9. $\frac{21 - 9}{3 + 1}$
10. $8^2 + 1$
11. $15 - \sqrt{9} \cdot 1.2$
12. $[(16 + 4) - 2.5] + 1^2$

Evaluate. Section 1.2

13. $14 - a^2$, if $a = 3$
14. $\sqrt{e} + 7.2$, if $e = 25$
15. $\frac{b}{2} + 10c$, if $b = 14$ and $c = 0.2$
16. $ab - 5c$, if $a = 2$, $b = 0.5$, and $c = 0.1$

Translate each phrase into an algebraic expression. Section 1.3

17. the difference of a number and 17
18. 23 multiplied by some number
19. the quotient of 6 and v
20. 13.5 more than a number

Translate each sentence into an equation. Section 1.3

21. Fifteen more than r is sixty-one.
22. The quotient of w and 6 is 78.
23. The product of 3.5 and a is 8.4.
24. Nineteen decreased by x is eight.

Replace each n with a number to make a true sentence. Sections 1.4, 1.8

25. $-42 \cdot 40 = n \cdot -42$
26. $5 + n = 5$
27. $n + \left(\frac{1}{3} + \frac{5}{6}\right) = \left(\frac{2}{3} + \frac{1}{3}\right) + \frac{5}{6}$
28. $n \cdot -729 = -729$
29. $n \cdot \frac{1}{8} = 1$
30. $n(3 + 1) = (4 \cdot 3) + (4 \cdot 1)$

State the opposite of each number. Section 1.5

31. -2   32. 5   33. 30   34. -6.7   35. $\frac{7}{8}$ 36. $-5\frac{3}{4}$

Find each value. Section 1.5

37. $|-9|$ 38. $|14|$ 39. $|-5| + |9|$ 40. $3|-3|$

Replace each ● with <, >, or = to make a true statement. Section 1.5

41. -4 ● 2   42. 3.5 ● -5.3   43. -1 ● 0   44. -2 ● $|-3|$

Compute. Sections 1.6–1.7

45. 72 − -25   46. -27 − 8   47. 37 + 13   48. 40 + -11
49. -9 • 14   50. -15 • -6   51. -66 ÷ -3   52. 34 ÷ 4
53. -31 + 30   54. -6 − -24   55. 2.13 ÷ -0.3   56. 9.36 + -4.59

Solve each equation. Sections 1.9–1.10

57. $15 = p + 11$ 58. $y + -2 = 18$ 59. $\frac{b}{4} = 20$ 60. $-4x = -72$

61. $x - 3 = -4$ 62. $-3x = 24$ 63. $56 = 8y$ 64. $\frac{p}{5} = -2$

65. $-6g = -516$ 66. $-36 = h - -12$ 67. $2a - 1 = 13$ 68. $\frac{c}{4} + 8 = 16$

69. $-4x + 7 = 32$ 70. $2t + 11 = -41$ 71. $8 - 5b = -17$ 72. $-6p + 14 = 32$

Translate each problem into an equation. Then solve the equation. Section 1.11

73. Twice g is twenty-five.

74. A number decreased by 40 is 23.

75. Jaoquin buys 3 dozen light bulbs. After changing the bulbs in his house, he has 15 bulbs left. How many light bulbs did he use?

76. Sally earns $5.00 for mowing the lawn and $4.00 for weeding the garden. How many times must she mow the lawn to earn $60.00?

77. Five less than three times a number is negative eleven.

78. Eight more than the quotient of a number and four is two.

79. Elena ordered packs of sunflower seeds. Each pack cost $6. She also paid $10 for shipping. If she spent a total of $82, how many packs did she order?

80. The Empire State Building in New York City is 1,250 feet tall. It has 103 floors. What is the height of each floor?

Chapter 1 Review

Chapter 1 Test

Evaluate.

1. $4 \cdot 3 + 2 \cdot 4$
2. $7 \cdot 6 - 12 \div 2$
3. $7^2 - (3 + \sqrt{4})$

Evaluate each expression if $x = 5$, $y = 4$, and $z = 0.5$.

4. $6z + 1$
5. $\dfrac{x + \sqrt{y}}{6}$
6. $x + y^2$
7. $(x + 4) \div (2 - z)$

Name the property shown.

8. $45 = 0 + 45$
9. $-4.2 + 7.5 = 7.5 + -4.2$
10. $(0.5 \cdot 5) \cdot 4 = 0.5 \cdot (5 \cdot 4)$
11. $3 \cdot (300 + 48) = (3 \cdot 300) + (3 \cdot 48)$

Find each value.

12. $|5|$
13. $|-17|$
14. $|-4.51|$
15. $\left|1\tfrac{1}{2}\right|$

Order the numbers in each list from least to greatest.

16. $-5, 14, 0, -10, 1.5$
17. $10, -15, -7.2, 3.1, -0.6$
18. $1, |-2|, -\tfrac{1}{2}, \tfrac{3}{4}, 0$

Compute.

19. $-16 + -11$
20. $-25 + 17$
21. $-5 - 18$
22. $23 - -16$
23. $14 \cdot -6$
24. $13 \cdot 27$
25. $312 \div 24$
26. $-72 \div -9$

Solve each equation. Check your solution.

27. $6 + x = 25$
28. $4.7 = t + 2.9$
29. $n - 12 = 42$
30. $1.8 = g - 3.4$
31. $9z = 36$
32. $1.3 = \dfrac{m}{7}$
33. $2t - 17 = 55$
34. $4.6 = \dfrac{d}{6} + 0.9$
35. $4x + 8 = 12$

Translate each sentence into an equation. Then solve each equation.

36. The product of 8 and t is 20.
37. Six increased by a is nineteen.
38. Four less than twice a number is seventy-eight.
39. The difference of 5 times a number and 7 is 45.

Solve.

40. A rental car service charges a fee of $10 plus $25 a day. Jason paid $160. For how many days did he rent the car?

41. Rosa bought two milk shakes and a hamburger. The hamburger cost $2, and the total bill was $8. How much did each milk shake cost?

Change of Pace

Modular Arithmetic

Assign each day of the week a number. Suppose today is day 2, or Tuesday. You can compute what day it will be 18 days from now as follows.

0—Sunday
1—Monday
2—Tuesday
3—Wednesday
4—Thursday
5—Friday
6—Saturday

A. Add 2 + 18.   **2 + 18 = 20**   The 2 represents Tuesday.
B. Divide 20 by 7.   **20 ÷ 7 = 2 R6**   There are 7 days in a week.
C. The remainder 6 means that 18 days after Tuesday is Saturday.

Computing in this way with the digits 0 through 6 is called **modulo 7** arithmetic. In modulo 7, the result is always 0, 1, 2, 3, 4, 5, or 6. Add 2 + 18 in modulo 7 as follows.

2 + 18   ≅   **6 (mod 7)**
2 plus 18   is equivalent to   6 modulo 7

Other examples are 7 + 14 ≅ 0 (mod 7) and 8 • 5 ≅ 5 (mod 7).

Copy and complete the following.

1. 5 + 23 ≅ ■ (mod 7)
2. 7 • 6 ≅ ■ (mod 7)
3. 1 + 3 ≅ ■ (mod 7)
4. 7 + 2 ≅ ■ (mod 7)
5. 8 • 5 ≅ ■ (mod 9)
6. 24 + 61 ≅ ■ (mod 12)
7. 9 • 12 ≅ ■ (mod 12)
8. 9 + 1 ≅ ■ (mod 6)
9. 26 • 26 ≅ ■ (mod 8)

Solve.

10. What number in modulo 7 is equivalent to 7?

11. What is an everyday use of modulo 12 arithmetic?

Use modular arithmetic to solve each problem.

12. Ines works 5 hours and goes home at 3:00 P.M. At what time did she start working?

13. Anton buys a clock on Monday. On what day of the week does the 30-day guarantee expire?

14. Julie buys a television on Friday. On what day of the week does the 90-day guarantee expire?

★15. Thomas Lynd has a 30-month installment auto loan. The first payment is due in October. In what month is the last payment due?

Cumulative Test

1. What is the value of $14 - 2 \cdot 3 + 4^2$?
   a. 12
   b. 24
   c. 52
   d. 228

2. Which expression can be written as $9 \cdot (12 + 7)$?
   a. $9 + (12 + 7)$
   b. $(9 + 12) \cdot (9 + 7)$
   c. $9 \cdot 12 + 7$
   d. $(9 \cdot 12) + (9 \cdot 7)$

3. How is the sentence "Six more than x is nine" written as an equation?
   a. $6 = x + 9$
   b. $6 + 9 = x$
   c. $6 + x = 9$
   d. none of the above

4. Brenda drove her car 1,048 miles in 8 months. What is the average distance she drove per month?
   a. 131 miles
   b. 1,056 miles
   c. 8,384 miles
   d. none of the above

5. Sam pays $1.89 for a sandwich and $0.55 for a drink. How much change should he receive from $10.00?
   a. $2.39
   b. $2.44
   c. $7.56
   d. $8.56

6. What property is shown by $2.1 + 5.03 = 5.03 + 2.1$?
   a. associative
   b. commutative
   c. distributive
   d. identity

7. The repair bill for John Salazar's car is $25.00 more than the mechanic's estimate. The total bill is $350.00. How can you find the amount of the mechanic's estimate?
   a. Multiply $350.00 by $25.00.
   b. Add $25.00 and $350.00.
   c. Subtract $25.00 from $350.00.
   d. none of the above

8. How many more televisions were sold on Monday than on Tuesday?
   a. 2
   b. 5
   c. 10
   d. 20

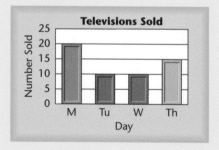

9. Ted works 5 hours a day for 4 days a week. He earns $4.30 an hour. Which problem *cannot* be solved using the information given?
   a. How much does Ted earn each week?
   b. How many days must Ted work to earn $500.00?
   c. How much more would Ted earn each week if he earns $4.75 an hour?
   d. What can Ted buy with the money he earns in 3 weeks?

10. The Smiths traveled 302 miles the first day of their vacation and 247 miles the next day. Anne Smith said they were 549 miles from home. Which fact did Anne assume?
    a. Each day they traveled in a straight line from home.
    b. The Smiths' car averaged 26 miles per gallon.
    c. They traveled faster the first day.
    d. none of the above

CHAPTER 2

Fractions and Exponents

Christian Hoehner
Virginia

2.1 Fractions and Decimals

Objective: to convert between fractions and decimals

Edgar's Jewelry Store has two diamonds on display. One weighs $\frac{3}{8}$ of a carat. The other diamond weighs $1\frac{1}{3}$ carats.

You can rename $\frac{3}{8}$ as a decimal by dividing.

$$\frac{3}{8} \rightarrow \begin{array}{r} 0.375 \\ 8\overline{)3.000} \\ \underline{-24} \\ 60 \\ \underline{-56} \\ 40 \\ \underline{-40} \\ 0 \end{array}$$

$\frac{3}{8} = 0.375$

Such a decimal is called a **terminating decimal**.

To rename $1\frac{1}{3}$ as a decimal, rename $\frac{1}{3}$ as a decimal, and then add the whole number.

$$\frac{1}{3} \rightarrow \begin{array}{r} 0.333 \\ 3\overline{)1.000} \end{array}$$

$1\frac{1}{3} = 1.33\ldots$ or $1.\overline{3}$.

A decimal, like $1.\overline{3}$, is called a **repeating decimal**. The bar is used to show which digit (or digits) repeats.

Numbers, such as $\frac{3}{8}$ and $1\frac{1}{3}$, are called **rational numbers**. Any number that can be represented as a quotient of two integers when the divisor is not 0, is called a rational number. All integers, fractions, and mixed numbers are rational numbers.

More Examples

Terminating decimals can be changed to fractions or mixed numbers as shown below.

A. $0.6 = \dfrac{\overset{3}{\cancel{6}}}{\underset{5}{\cancel{10}}}$

$= \dfrac{3}{5}$

B. $3.16 = 3\dfrac{\overset{4}{\cancel{16}}}{\underset{25}{\cancel{100}}}$

$= 3\dfrac{4}{25}$

Repeating decimals can be written as fractions as shown below.

C. Express $0.6666\ldots$, or $0.\overline{6}$, as a fraction.

Let $x = 0.\overline{6}$ or $0.6666\ldots$.

$10x = 6.\overline{6}$ Multiply both sides by 10 because there is 1 repeating digit.

$\underline{-x = -0.\overline{6}}$

$\dfrac{9x}{9} = \dfrac{6}{9}$ Then subtract x from $10x$ and $0.\overline{6}$ from $6.\overline{6}$.

$x = \dfrac{6}{9} = \dfrac{2}{3}$ Solve for x. Simplify.

The decimal $0.\overline{6}$ is equal to the fraction $\frac{2}{3}$.

Try THESE

Write T if the decimal is terminating or R if the decimal is repeating. Write each repeating decimal using bar notation.

1. 0.375
2. 0.444 . . .
3. 0.2727 . . .
4. $\frac{5}{6}$ 6)5.000
5. $\frac{9}{16}$ 16)9.000

Find the value of each variable.

6. $0.23 = \frac{x}{100}$
7. $4.9 = 4\frac{a}{10}$
8. $0.077 = \frac{77}{t}$
9. $0.50 = \frac{r}{100}$

Exercises

Determine whether each statement is *true* or *false*.

1. $\frac{7}{8} = 0.875$
2. $\frac{15}{4} = 3.\overline{7}$
3. $\frac{5}{13} = 0.\overline{38}$
4. $2\frac{1}{12} = 2.08\overline{3}$

Rename each fraction or mixed number as a decimal. Use bar notation for repeating decimals.

5. $\frac{6}{11}$
6. $\frac{15}{8}$
7. $\frac{5}{9}$
8. $1\frac{13}{25}$
9. $\frac{22}{45}$
10. $3\frac{21}{40}$
11. $6\frac{5}{12}$
12. $\frac{109}{30}$

Rename each decimal as a fraction or mixed number in simplest form.

13. 0.69
14. 0.5
15. 0.55
16. 0.15
17. 0.250
18. 0.065
19. 0.437
20. 1.10
21. 1.25
22. 1.06
23. 3.125
24. 4.050

Rename the following decimals as fractions in simplest form.

25. $0.\overline{8}$
26. $1.\overline{74}$
27. $0.\overline{018}$
28. $0.1\overline{63}$

Problem SOLVING

29. Write an example of a repeating decimal that is between 9.16 and 9.17. Use bar notation.

Constructed RESPONSE

30. If $3 + \frac{1}{4}$ can be written as the improper fraction $\frac{13}{4}$, how would you write $n + \frac{1}{4}$ as an improper fraction? Explain your reasoning using your knowledge of fractions.

31. Batting averages are usually expressed as decimals. Joe had 42 hits in 120 at bats. Damien had 35 hits in 105 at bats.

    a. Find Joe's and Damien's batting average.

    b. Based on their batting averages, who is more likely to get a hit?

2.1 Fractions and Decimals

2.2 Comparing and Ordering Rational Numbers

Objective: to compare and order rational numbers

Carla and Ken are debating about who bought more oranges. Carla bought $\frac{5}{8}$ of a pound, and Ken bought $\frac{3}{5}$ of a pound. Who bought more oranges?

You can use two methods to compare the rational numbers $\frac{5}{8}$ and $\frac{3}{5}$.

Common Denominator Method	Decimal Method
Write the fractions with the same denominator. The least common denominator of $\frac{5}{8}$ and $\frac{3}{5}$ is 40. $$\frac{5}{8} = \frac{5 \cdot 5}{8 \cdot 5} = \frac{25}{40} \qquad \frac{3}{5} = \frac{3 \cdot 8}{5 \cdot 8} = \frac{24}{40}$$ Since $\frac{25}{40} > \frac{24}{40}, \frac{5}{8} > \frac{3}{5}$.	Convert the fractions to decimals. $$5 \div 8 = 0.625 \qquad 3 \div 5 = 0.6$$ $$\frac{5}{8} = 0.625 \qquad \frac{3}{5} = 0.6$$ Since $0.625 > 0.6$, $\frac{5}{8} > \frac{3}{5}$.

Carla bought more oranges than Ken.

Examples

A. Order $\frac{2}{5}$, 0.367, 0.376, and $\frac{3}{8}$ from least to greatest.

1. Convert the fractions to decimals. $\frac{2}{5} = 0.4$ $\frac{3}{8} = 0.375$
2. Order the decimals.   0.367, 0.375, 0.376, 0.4

The order from least to greatest is 0.367, $\frac{3}{8}$, 0.376, and $\frac{2}{5}$.

B. Compare $-6\frac{1}{4}$ and -6.20.

Convert the fraction to a decimal. $\frac{1}{4} = 0.25$ so $-6\frac{1}{4} = -6.25$

Since $-6.20 > -6.25$, $-6.20 > -6\frac{1}{4}$.

> On a number line, a number is greater than any number to its left.

Check your answer using a number line.

Try THESE

Replace each ■ with <, >, or = to make a true statement.

1. 2.5 ■ $2\frac{1}{2}$
2. -0.4 ■ -0.38
3. $\frac{3}{2}$ ■ $\frac{-7}{2}$
4. $\frac{3}{4}$ ■ $\frac{7}{12}$

Exercises

Replace each ■ with <, >, or = to make a true statement.

1. -4.6 ■ 4.6
2. -0.8 ■ -0.7
3. $\frac{1}{2}$ ■ -1
4. $-1\frac{1}{4}$ ■ $-3\frac{3}{4}$
5. $\frac{4}{5}$ ■ $\frac{3}{5}$
6. $\frac{3}{5}$ ■ $\frac{6}{10}$
7. $\frac{5}{2}$ ■ 0.4
8. $\frac{2}{3}$ ■ 0.67
9. |-5| ■ |-3.5|
10. 6.3 ■ $6\frac{5}{16}$
11. $\frac{-4}{5}$ ■ $\frac{-8}{9}$
12. $\frac{5}{7}$ ■ 0.63
13. $\frac{3}{8}$ ■ 0.375
14. $\frac{-4}{9}$ ■ $\frac{5}{8}$
15. $\frac{2}{4}$ ■ 0.05
★16. $0.\overline{25}$ ■ $\frac{1}{4}$

Order the numbers in each list from least to greatest.

17. -1.5, -1.8, -1.4, -1.1
18. -3.142, -3.51, -3.211
19. 0, $\frac{-1}{2}$, 0.45, -0.55
20. 0.3, $\frac{1}{5}$, 0.4, $\frac{1}{8}$
21. $\frac{22}{7}$, 3.14, 3.144, $\frac{20}{6}$
22. $4\frac{1}{6}$, 4.28, 4.55, $4\frac{2}{7}$

Problem SOLVING

23. Ester sprinted 182.31 meters in 20.6 seconds. Estimate her speed per second.

★24. Change $\frac{2\frac{2}{3}}{3\frac{1}{2}}$ to a rational number in simplest form.

Constructed RESPONSE

25. Is $\frac{65}{100}$ equal to $0.\overline{65}$? Explain why or why not.

26. If 1 is added to the numerator and denominator of the fraction $\frac{3}{4}$, is the new fraction greater than or less than the original fraction? Explain.

Test PREP

27. Determine which statement is true.
    a. -0.65 > -0.60
    b. $\frac{2}{3}$ > 0.75
    c. 0.75 = $\frac{3}{4}$
    d. $\frac{-3}{5}$ < -0.60

28. Which set of numbers is in order from least to greatest?
    a. $\frac{1}{4}$, $\frac{4}{10}$, 0.04, 0.44
    b. $\frac{4}{10}$, $\frac{1}{4}$, 0.44, 0.04
    c. 0.44, $\frac{4}{10}$, $\frac{1}{4}$, 0.04
    d. 0.04, $\frac{1}{4}$, $\frac{4}{10}$, 0.44

2.3 Adding and Subtracting Fractions and Mixed Numbers

Objective: to add and subtract fractions and mixed numbers

Jennifer de Casas plants two sizes of marigolds. One marigold is $2\frac{3}{4}$ inches taller than the other. The shorter marigold is $3\frac{7}{8}$ inches tall.

To find the height of the taller marigold, add $2\frac{3}{4}$ and $3\frac{7}{8}$.   Estimate.
$$3 + 4 = 7$$

Add with mixed numbers as follows.

Step 1	Step 2
Find the LCD. Rename each fraction. $$2\frac{3}{4} = 2\frac{6}{8}$$ $$+\ 3\frac{7}{8} = 3\frac{7}{8}$$	Add and simplify, if necessary. $$2\frac{3}{4} = 2\frac{6}{8}$$ $$+\ 3\frac{7}{8} = 3\frac{7}{8}$$ $$5\frac{13}{8} = 6\frac{5}{8}$$

The taller marigold is $6\frac{5}{8}$ inches tall.   Is the answer close to the estimate?

More Examples

A. $\frac{-3}{8} + \frac{7}{8} = \frac{-3 + 7}{8}$ Add the numerators.

$\phantom{\frac{-3}{8} + \frac{7}{8}} = \frac{4}{8}$ Keep the denominators the same.

$\phantom{\frac{-3}{8} + \frac{7}{8}} = \frac{4}{8}$ or $\frac{1}{2}$

B. $-4\frac{5}{6} = -4\frac{5}{6}$ Estimate. $-5 + -6 = -11$

$+\ -5\frac{2}{3} = -5\frac{4}{6}$

$-9\frac{9}{6} = -10\frac{3}{6}$ or $-10\frac{1}{2}$

Write each sum in simplest form.

Using a least common denominator can also help you subtract fractions.
Find $5\frac{1}{3} - 2\frac{1}{2}$.

Step 1	Step 2	Step 3
Find the LCD. Rename each fraction.	Rename, if necessary.	Subtract.
$5\frac{1}{3} = 5\frac{2}{6}$ $-2\frac{1}{2} = 2\frac{3}{6}$	$5\frac{1}{3} = 5\frac{2}{6} = 4\frac{8}{6}$ $-2\frac{1}{2} = 2\frac{3}{6} = 2\frac{3}{6}$	$5\frac{1}{3} = 5\frac{2}{6} = 4\frac{8}{6}$ $-2\frac{1}{2} = 2\frac{3}{6} = 2\frac{3}{6}$ $2\frac{5}{6}$

More Examples

C. $5\frac{1}{4} = 5\frac{5}{20} = 4\frac{25}{20}$
$-2\frac{3}{5} = 2\frac{12}{20} = 2\frac{12}{20}$
$\phantom{-2\frac{3}{5} = 2\frac{12}{20} = {}} 2\frac{13}{20}$

D. $6 = 5\frac{6}{6}$
$-3\frac{1}{6} = 3\frac{1}{6}$
$\phantom{-3\frac{1}{6} = {}} 2\frac{5}{6}$

▶ Rename the whole number.

E. $4\frac{3}{8} - 2 = 2\frac{3}{8}$

Try THESE

Rename.

1. $4 = 3\frac{\blacksquare}{5}$
2. $10 = 9\frac{\blacksquare}{8}$
3. $5\frac{1}{2} = 4\frac{\blacksquare}{2}$
4. $3\frac{1}{3} = 3\frac{2}{6} = 2\frac{\blacksquare}{6}$

Subtract. Write each difference in simplest form.

5. $\frac{5}{7} - \frac{3}{7}$
6. $9\frac{2}{9} - 6\frac{4}{9}$
7. $4\frac{5}{8} \quad 4\frac{5}{8}$
   $-2\frac{1}{4} \quad -2\frac{2}{8}$
8. $10\frac{3}{10} \quad 10\frac{6}{20} = 9\frac{26}{20}$
   $-5\frac{3}{4} \quad -5\frac{15}{20} = 5\frac{15}{20}$

Exercises

Add or subtract. Write your answer in simplest form.

1. $\frac{-3}{10} + \frac{-1}{2}$
2. $\frac{1}{3} + \frac{1}{2}$
3. $-5\frac{1}{6} + 7\frac{1}{3}$
4. $-10\frac{1}{3} + \left(-6\frac{1}{4}\right)$
5. $-7\frac{5}{7} - 4\frac{3}{7}$
6. $\frac{11}{12} - \frac{1}{6}$
7. $17\frac{11}{12} - 13\frac{1}{6}$
8. $-17\frac{1}{3} - 9\frac{1}{2}$
9. $8\frac{3}{8} + 9\frac{1}{3}$

10. $11 - 3\frac{5}{9}$
11. $12\frac{1}{2} - 8\frac{2}{3}$
12. $11\frac{1}{3} + 5\frac{2}{5}$
13. $-14\frac{5}{8} - 6\frac{5}{6}$
14. $-14 - 9\frac{3}{5}$
15. $22\frac{5}{9} - 21\frac{5}{6}$
16. $2\frac{1}{3} + 3\frac{1}{4} + 4\frac{1}{6}$
17. $6\frac{7}{8} + 9 + 7\frac{3}{4}$
18. $3\frac{4}{5} + 8\frac{1}{4} + 6\frac{5}{6}$

Evaluate each expression if $a = \frac{1}{3}$, $b = \frac{5}{6}$, $c = 2\frac{3}{4}$, and $d = 1\frac{2}{3}$.

19. $a + b$
20. $c - a$
21. $(d + a) - b$
22. $c - (b - a)$

Problem SOLVING

Solve. Use the diagram to answer questions 23–25.

23. How thick is the insulation and the drywall?
24. How thick are the outside layers of the wall including the wall sheathing and siding?
25. How thick is the entire wall?
26. There were three snowstorms last winter. The snow falls measured $3\frac{1}{8}$ in., $12\frac{4}{5}$ in., and $5\frac{1}{4}$ in. How much snow fell in total?

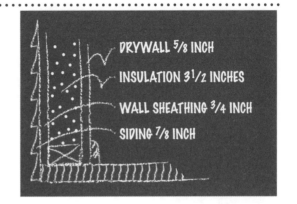

DRYWALL ⅝ INCH
INSULATION 3½ INCHES
WALL SHEATHING ¾ INCH
SIDING ⅞ INCH

27. A dog had two puppies. The puppies weighed $5\frac{1}{5}$ lb and $4\frac{1}{2}$ lb. What is the difference in the weights of the puppies?

Constructed RESPONSE

28. Draw a picture of $\frac{3}{4} + \frac{1}{3}$. Then write directions for a sixth-grade student explaining how to add two fractions. Tell why you add fractions the way you do.

Mind BUILDER

Adding Algebraic Fractions

To add algebraic fractions, use the same methods as when you add common fractions.

Find $\frac{2}{x} + \frac{3}{x}$.

$\frac{2}{x} + \frac{3}{x} = \frac{2+3}{x} = \frac{5}{x}$

Find $4\frac{5}{y} + \frac{3}{y}$.

$4\frac{5}{y} + \frac{3}{y} = 4 + \frac{5+3}{y} = 4\frac{8}{y}$

Find each sum.

1. $\frac{4}{a} + \frac{3}{a}$
2. $\frac{b}{x} + \frac{2}{x}$
3. $\frac{4k}{t} + \frac{6k}{t}$
4. $\frac{2n}{m} + \frac{3n}{m}$

Problem Solving

Whodunit?

Some priceless jewels have been stolen from the Smythe Family Mansion. The famous detective I. L. Findit has narrowed the list of suspects to four: the maid, the butler, the ne'er-do-well nephew, and the famous daughter.

When questioned, each suspect made one statement and refused to say more.

- The maid said, "I did not do it."
- The ne'er-do-well nephew said, "The famous daughter did it."
- The famous daughter said, "The butler did it."
- The butler said, "The famous daughter lied when she said I did it."

Assume, in turn, that each person committed the theft. Use the assumptions to discover which statements might be true. For example, if the butler did it, then his statement and the nephew's statement are false, and the maid and daughter's statements are true. Organize the assumptions and true and false statements in a grid to answer the questions.

1. Suppose one statement is true.
   Who stole the jewels?

2. Suppose one and only one statement is false.
   Who stole the jewels?

Extension

Can two and only two statements be true? Explain why or why not.

2.4 Multiplying and Dividing Fractions and Mixed Numbers

Objective: to multiply and divide fractions and mixed numbers

A wax Begonia grows $1\frac{1}{6}$ feet tall. The heights of other houseplants are given in relation to the wax Begonia.

Dracaena

Boston fern

Coleus

African violet   wax Begonia

$\frac{3}{7}$ $1\frac{5}{7}$ $2\frac{1}{7}$ $4\frac{7}{8}$

A Dracaena grows $4\frac{7}{8}$ times as tall as a wax Begonia. To find the height of a Dracaena, multiply $4\frac{7}{8}$ and $1\frac{1}{6}$.

Estimate. $5 \cdot 1 = 5$

$4\frac{7}{8} \cdot 1\frac{1}{6} = \frac{39}{8} \cdot \frac{7}{6}$ Rename the mixed numbers as fractions.

$= \frac{\overset{13}{\cancel{39}}}{8} \cdot \frac{7}{\underset{2}{\cancel{6}}}$ The GCF of 39 and 6 is 3, so use this shortcut.

$= \frac{91}{16}$ $91 \div 16 = 5$ R11 or $5\frac{11}{16}$

$= 5\frac{11}{16}$ The estimate shows that the answer is reasonable.

The Dracaena is $5\frac{11}{16}$ feet tall.

More Examples

A. $\frac{1}{4} \cdot \frac{2}{3} = \frac{1 \cdot 2}{4 \cdot 3}$ Multiply the numerators and denominators.

$= \frac{\overset{1}{\cancel{2}}}{\underset{6}{\cancel{12}}}$ The GCF of 2 and 12 is 2.

$= \frac{1}{6}$

B. $\frac{5}{6} \cdot 18 = \frac{5}{6} \cdot \frac{18}{1}$ Rename the whole number as a fraction.

$= \frac{5 \cdot \overset{3}{\cancel{18}}}{\underset{1}{\cancel{6}} \cdot 1}$ The GCF of 18 and 6 is 6.

$= \frac{15}{1}$ or 15

The number $\frac{1}{2}$ is called the **multiplicative inverse** of 2. Numbers, like $\frac{1}{2}$ and 2, that are multiplicative inverses of each other are called **reciprocals**. Reciprocals are two numbers whose product is 1.

$\frac{1}{8}$ and 8 are reciprocals because $\frac{1}{8} \cdot 8 = 1$.

$\frac{2}{3}$ and $\frac{3}{2}$ are reciprocals because $\frac{2}{3} \cdot \frac{3}{2} = 1$.

To divide fractions, rewrite the division as a related multiplication problem in which you multiply the reciprocal of the divisor.

C. $10\frac{1}{2} \div 1\frac{3}{4} = \frac{21}{2} \div \frac{7}{4}$ Rename the mixed numbers as fractions.

$= \frac{21}{2} \cdot \frac{4}{7}$ Multiply by the reciprocal of $\frac{7}{4}$.

$= \frac{\overset{3}{\cancel{21}}}{\underset{1}{\cancel{2}}} \cdot \frac{\overset{2}{\cancel{4}}}{\underset{1}{\cancel{7}}}$ The GCF of 21 and 7 is 7.

$= \frac{6}{1}$ or 6   The GCF of 4 and 2 is 2.

D. $\frac{3}{4} \div \frac{5}{8} = \frac{3}{4} \cdot \frac{8}{5}$

$= \frac{3}{\underset{1}{\cancel{4}}} \cdot \frac{\overset{2}{\cancel{8}}}{5}$ Multiply by the reciprocal of $\frac{5}{8}$.

$= \frac{6}{5}$ or $1\frac{1}{5}$

Try THESE

State if the quotient will be *less than* or *greater than* 1.

1. $2\frac{1}{3} \div 4$
2. $5\frac{1}{2} \div 3\frac{1}{4}$
3. $13 \div 4\frac{1}{2}$
4. $3\frac{3}{5} \div 6$
5. $\frac{3}{4} \div \frac{1}{4}$
6. $\frac{6}{7} \div 6$
7. $\frac{5}{6} \div \frac{5}{12}$
8. $2\frac{1}{4} \div 1\frac{7}{8}$

Exercises

Multiply. Write each product in simplest form.

1. $\frac{3}{4} \cdot \frac{2}{5}$
2. $\frac{-3}{4} \cdot \frac{-3}{5}$
3. $2\frac{1}{2} \cdot \frac{2}{3}$
4. $\frac{3}{11} \cdot -3\frac{2}{3}$
5. $9 \cdot 4\frac{2}{3}$
6. $5\frac{3}{7} \cdot 14$
7. $6 \cdot \frac{3}{5}$
8. $\frac{1}{4} \cdot 12$
9. $-9\frac{1}{2} \cdot 4\frac{1}{2}$
10. $6\frac{1}{4} \cdot 2\frac{2}{3}$
11. $-2\frac{4}{7} \cdot 1\frac{5}{16}$
12. $-2\frac{7}{10} \cdot -2\frac{1}{12}$

★13. $\frac{5}{6} \cdot \frac{21}{25} \cdot \frac{2}{7}$
★14. $1\frac{3}{8} \cdot 2 \cdot \frac{4}{11}$
★15. $3\frac{4}{5} \cdot 1\frac{7}{8} \cdot 2\frac{7}{19}$

16. What fraction can replace n to make $\frac{2}{3} \cdot n = \frac{3}{2}$ a true sentence?
17. What fraction can replace n to make $\frac{2}{3} \cdot n = 78$ a true sentence?

Divide. Write each quotient in simplest form.

18. $\frac{5}{7} \div \frac{1}{7}$
19. $\frac{-4}{9} \div \frac{1}{3}$
20. $\frac{8}{9} \div 4$
21. $9 \div \frac{3}{8}$
22. $\frac{-4}{9} \div \frac{-8}{21}$
23. $1\frac{3}{5} \div \frac{5}{8}$
24. $\frac{8}{9} \div -2\frac{2}{5}$
25. $\frac{5}{6} \div 1\frac{1}{9}$
26. $-1\frac{2}{5} \div -2\frac{2}{3}$
27. $3\frac{1}{5} \div -\frac{1}{3}$

2.4 Multiplying and Dividing Fractions and Mixed Numbers

Evaluate each expression if $a = 1\frac{1}{2}$, $b = \frac{3}{4}$, and $c = 3$.

28. $a \div b$
29. $a \div c^2$
30. $\frac{c}{b}$
31. $\frac{ac}{b}$
★32. $\frac{(a-b)}{c}$

Problem SOLVING

33. Mr. Valdez had $2\frac{1}{2}$ dozen golf balls. He lost $\frac{1}{3}$ of them. How many golf balls does he have left?

34. Ayita is $\frac{2}{3}$ as old as Kameko. The difference in their ages is 4 years. How old is each girl now?

35. The perimeter of a three-sided garden is 36 feet. Side **A** is $\frac{3}{4}$ the length of side **B**. Side **C** is 15 feet long. What are the lengths of side **A** and side **B**?

36. A $\frac{3}{4}$-pound bag of dried fruit costs $1.89. What is the price per pound? What is the price per ounce? (16 ounces = 1 pound)

37. Eloise worked 8, $7\frac{1}{2}$, $6\frac{1}{4}$, and $4\frac{3}{4}$ hours on 4 weekends. Find the average number of hours she worked.

38. Describe the reciprocal of a number greater than 1. Describe the reciprocal of a number less than 1. Give examples.

Mind BUILDER

Dividing Algebraic Fractions

To divide algebraic fractions, use the same methods as when you divide common fractions.

$$2 \div \frac{1}{y} = \frac{2}{1} \div \frac{1}{y}$$
$$= \frac{2}{1} \cdot \frac{y}{1}$$
$$= \frac{2y}{1} \text{ or } 2y$$

$$\frac{5}{a} \div \frac{3}{a} = \frac{5}{a} \cdot \frac{a}{3}$$
$$= \frac{5 \cdot \overset{1}{\cancel{a}}}{\underset{1}{\cancel{a}} \cdot 3}$$
$$= \frac{5}{3} \text{ or } 1\frac{2}{3}$$

Find each quotient.

1. $\frac{3}{b} \div \frac{4}{a}$
2. $\frac{1}{x} \div \frac{m}{x}$
3. $\frac{5}{x} \div \frac{y}{3}$
4. $\frac{2n}{m} \div \frac{3}{2m}$

2.4 Multiplying and Dividing Fractions and Mixed Numbers

Cumulative Review

Compute mentally. Write only your answers.

1. 4,672
  + 298

2. 3.6
  − 2.58

3. $9.50
  × 2

4. $18\overline{)360}$

5. $5.00 − $3.69

6. 21 + 62

7. 18 • 1.5

8. $6.70 ÷ 10

Replace each n with a number to make a true sentence.

9. $n + (6 + 8) = (4 + 6) + 8$

10. $2.3 • 7 = n • 2.3$

11. $(7 • 32) + (7 • n) = 7 • (32 + 9)$

12. $(3 • 5) • 16 = 3 • (n • 16)$

Evaluate each expression if $a = 4$ and $b = 0.5$.

13. $a + b$

14. $3a$

15. $4a + 5b$

16. $(a − b^2) • b$

17. $3 • (2a + b)$

18. $(a + b)^2$

19. $7 • (\sqrt{a} − 2b)$

Solve each equation. Check your solution.

20. $10 + c = 16$

21. $n − 9 = 12$

22. $9y = 8.1$

23. $8 = \frac{a}{4}$

24. $2.4 = s − 0.75$

25. $a + 3.2 = 7.3$

26. $0.35b = 7$

27. $\frac{g}{1.5} = 4$

28. $2x − 4 = 18$

Find the greatest common factor (GCF) of each group of numbers.

29. 24, 9

30. 18, 12

31. 15, 8

32. 12, 18, 20

Find the least common multiple (LCM) of each group of numbers.

33. 3, 2

34. 8, 16

35. 10, 4

36. 4, 6, 5

Solve.

37. After three tests, Lacy has an average of 86. What score must she receive on a fourth test in order to have an average of at least 88?

38. Calhoun County in Alabama has a population of 116,936 and an area of 611 square miles. Estimate the population per square mile.

39. Danny Sullivan drove a lap at the Indianapolis 500 in 41.5 seconds. One lap is 13,200 feet. Find the speed to the nearest tenth of a foot per second.

40. Mr. Devine pays $1.34 a gallon for gasoline. The gasoline tank in his car holds 14.5 gallons. His car travels 27 miles per gallon. Can he drive for 400 miles on one tank of gasoline?

2.5 Solving Equations with Fractions and Mixed Numbers

Objective: to solve equations with fractions and mixed numbers

Mike O'Reilly mows grass for $\frac{3}{4}$ of an hour. That is $\frac{1}{3}$ of an hour more than the time he spends pulling weeds. To find how long Mike takes to pull the weeds, you can write an equation.

The sum of the weeding time and $\frac{1}{3}$ of an hour equals the mowing time.

$$w + \frac{1}{3} = \frac{3}{4}$$

Equations involving fractions and mixed numbers are solved the same way as those involving whole numbers and decimals.

$$w + \frac{1}{3} = \frac{3}{4}$$

$w + \frac{1}{3} - \frac{1}{3} = \frac{3}{4} - \frac{1}{3}$ Subtract $\frac{1}{3}$ from each side.

$w = \frac{9}{12} - \frac{4}{12}$ Rename the fractions using the LCD.

$w = \frac{5}{12}$ of an hour or 25 minutes

Check: $w + \frac{1}{3} = \frac{3}{4}$

$\frac{5}{12} + \frac{1}{3} = \frac{3}{4}$ Replace w with $\frac{5}{12}$.

$\frac{5}{12} + \frac{4}{12} = \frac{9}{12}$

$\frac{9}{12} = \frac{9}{12}$ ✓

More Examples

A. $x - \frac{9}{14} = 2\frac{3}{7}$

$x - \frac{9}{14} + \frac{9}{14} = 2\frac{3}{7} + \frac{9}{14}$ Adding $\frac{9}{14}$ is the inverse of subtracting $\frac{9}{14}$.

$x = 2\frac{6}{14} + \frac{9}{14}$

$x = 2\frac{15}{14}$ or $3\frac{1}{14}$

Check: $3\frac{1}{14} - \frac{9}{14} = 2\frac{3}{7}$ ✓

B. $1\frac{1}{2}s = \frac{1}{2}$

$\frac{1\frac{1}{2}s}{1\frac{1}{2}} = \frac{\frac{1}{2}}{1\frac{1}{2}}$ Divide each side by $1\frac{1}{2}$.

$s = \frac{1}{2} \div \frac{3}{2}$

$s = \frac{1}{2} \cdot \frac{2}{3}$

$s = \frac{1}{3}$ Check: $1\frac{1}{2} \cdot \frac{1}{3} = \frac{1}{2}$ ✓

C. $\frac{n}{15} = 2\frac{3}{5}$ This could also be written $n \div 15 = 2\frac{3}{5}$.

$\frac{n}{15} \cdot \frac{15}{1} = 2\frac{3}{5} \cdot \frac{15}{1}$ Multiply by the inverse of $\frac{1}{15}$.

$\frac{n}{15} \cdot \frac{15}{1} = \frac{13}{5} \cdot \frac{15}{1}$

$n = 39$ Check: $\frac{39}{15} = 2\frac{3}{5}$ ✓

Try THESE

Name the operation and the number you would use to solve each equation.

1. $x + \frac{1}{4} = \frac{3}{4}$
2. $a - 1\frac{1}{2} = \frac{3}{8}$
3. $3y = 4\frac{1}{2}$
4. $\frac{t}{5} = 2\frac{1}{3}$

Exercises

Solve each equation. Check your solution.

1. $p + 1\frac{1}{2} = -3$
2. $b - 3\frac{1}{8} = 4\frac{3}{8}$
3. $r \cdot -1\frac{1}{3} = 4$
4. $\frac{y}{6} = 1\frac{1}{6}$
5. $7r = 3\frac{1}{2}$
6. $s + 1\frac{2}{3} = 2\frac{5}{6}$
7. $c - 4\frac{1}{2} = 1\frac{5}{8}$
8. $\frac{b}{10} = -2\frac{4}{5}$
9. $q + 2\frac{3}{4} = -3\frac{7}{8}$
10. $3n = \frac{3}{4}$
11. $s - -1\frac{1}{3} = 2\frac{1}{6}$
12. $\frac{t}{5} = 3\frac{2}{5}$
13. $a \div \frac{-1}{2} = 2\frac{1}{3}$
14. $9t = 4\frac{1}{2}$
15. $r + -5\frac{7}{8} = 9\frac{3}{8}$
16. $p - 1\frac{1}{4} = 1\frac{7}{8}$
17. $w \div -5 = 2\frac{5}{9}$
18. $-3q = 1\frac{2}{3}$

Translate each sentence into an equation. Then solve each equation.

19. The sum of some number and $\frac{2}{3}$ equals $3\frac{1}{3}$.
20. The product of $\frac{1}{5}$ and n equals $1\frac{2}{5}$.
21. The difference of x and $1\frac{7}{10}$ is equal to $4\frac{3}{15}$.
22. Some number divided by 9 is $2\frac{1}{6}$.

Problem SOLVING

Write an equation. Then solve.

23. A coat is on sale for $24.50. The sale price is $\frac{2}{3}$ of the regular price. What was the regular price?

24. Stock **XYZ** closed at $27\frac{5}{8}$ on Friday. That was a gain of $1\frac{7}{8}$ from the opening price. At what price did it open?

★ 25. Mr. Brophy's total bill at the grocery store is $33.50, but he uses several coupons and pays only $26.80. What fraction of the total bill did he actually pay?

★ 26. Study the pattern. Write an expression you can use to quickly find the sum of the first n odd numbers. Use the expression to find the sum of the first 65 odd numbers.

$1 + 3 = 4$
$1 + 3 + 5 = 9$
$1 + 3 + 5 + 7 = 16$

Mixed REVIEW

Compute.

27. $26.65 + 0.123 + 10.605$
28. $114.58 - $19.19
29. $0.18 + $10.04 + $27
30. $55 - 5.49$

2.6 Problem-Solving Strategy: Solve a Simpler Problem

Objective: to solve a problem by solving a simpler problem

The Royalton Garden Club takes care of a square flower bed in the park. They plant two different kinds of flowers in a checkerboard pattern. How many squares are there?

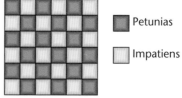
Petunias
Impatiens

1. READ The flower bed is a 6 × 6 square with two types of flowers alternating each way. You need to find the total number of squares of all sizes.

2. PLAN Find the number of squares in a 1 × 1 square, a 2 × 2 square, and a 3 × 3 square. Look for a pattern that you can extend to a 6 × 6 square.

3. SOLVE

1 × 1 square   2 × 2 square   3 × 3 square

1 × 1s: 1

1 × 1s: 4
+ 2 × 2s: 1
————
5

1 × 1s: 9
2 × 2s: 4
+ 3 × 3s: 1
————
14

Notice that the number of squares is the sum of the numbers squared up to that point. For a 6 × 6 square, there would be 1 + 4 + 9 + 16 + 25 + 36 or a total of 91 squares.

$1^2 + 2^2 + 3^2 + 4^2 + 5^2 + 6^2 = 91$

4. CHECK You can check the solution by actually continuing to list the number of squares for each size up to a 6 × 6 square. The answer is correct.

Solve. First solve a simpler problem.

1. How many numbers from 10 to 1,000 read the same forward or backward?

2. Suppose you add the digits of a number. How many numbers from 0 to 1,000 have a sum of 10?

Solve

1. How many triangles are there in the figure?

2. Find the sum of the numbers 1 through 1,000.

3. The Royalton Express seats 132 passengers. On a recent trip, there was one empty seat for every 3 passengers. How many passengers went on the trip?

4. Mrs. Trimmer grades a 20-question test. Each correct answer is +2 points. Each wrong answer is -1 point. Lana's score is 25. How many of her answers are correct?

5. Jose cuts a bolt of material that is 50 yards long into pieces that are 1 yard long. Each cut takes 1 minute. How long will it take Jose to make all the cuts?

6. A loaf of bread is 10 inches long. How many cuts are necessary to cut it into 20 slices?

7. In a basketball tournament, each team plays a game against each of the other teams. There are 10 teams. How many games will be played in the first round?

8. What is the largest number of pieces of pizza you can get with 6 straight cuts?

Mid-Chapter REVIEW

Rename each fraction or mixed number as a decimal.

1. $2\frac{3}{5}$
2. $\frac{12}{15}$
3. $5\frac{7}{9}$
4. $\frac{22}{40}$

Order the numbers in each list from least to greatest.

5. $0.88, \frac{4}{5}, \frac{3}{4}, 0.79$
6. $-0.58, \frac{-2}{3}, -0.62, \frac{-3}{5}$

Add or subtract. Write each answer in simplest form.

7. $\frac{7}{9} - \frac{1}{3}$
8. $\frac{11}{12} + \frac{-3}{8}$
9. $12 - 5\frac{3}{7}$
10. $-13\frac{5}{6} + -8\frac{3}{8}$

Multiply or divide. Write each answer in simplest form.

11. $\frac{-3}{4} \div 1\frac{1}{2}$
12. $\frac{5}{6} \cdot \frac{3}{8}$
13. $-2\frac{2}{5} \cdot -5\frac{1}{3}$
14. $3\frac{2}{3} \div -1\frac{1}{2}$

2.7 Exponents

Objective: to use powers and exponents in expressions

Exploration Exercise

1. Use your calculator to find each of the following.

    a. 4 • 4 • 4   b. 7 • 7 • 7 • 7   c. 2 • 2 • 2 • 2 • 2

2. You can find the value of 4 • 4 • 4, or 4^3, by pressing 4 x^y 3 on your scientific calculator. Find the value of each of the following.

    a. 4^3 b. 7^4 c. 2^5

3. What do you notice about the answers to problems 1–2?

An expression, such as 4 • 4 • 4, can be written as 4^3, which is read "4 to the third power."

$$\text{base} \longrightarrow 4^3 \longleftarrow \text{exponent}$$

The raised number 3 shows the number of times 4 is used as a factor. The number 4 is the **base**. The number 3 is the **exponent**. An expression using a base and an exponent is a **power**.

Factored Form	**Exponential Form**	**Standard Form**
4 • 4 • 4	4^3	64

A number with an exponent of 1 is the number itself. $8^1 = 8$

Examples

A. Write $x • x • x • y • y$ using exponents.

x^3y^2 x is a factor 3 times, and y is a factor 2 times.

B. Write -5 • -5 • -5 • -5 • $a • a • a$ using exponents.

$(-5)^4 a^3$ -5 is a factor 4 times, and a is a factor 3 times.

C. Evaluate. 6^3

$6^3 = 6 • 6 • 6 = 216$

Try THESE

Write in factored form and then in standard form.

1. 6^2 2. 2^4 3. 5 cubed   4. 3^3

Exercises

Write each expression using exponents.

1. $2 \cdot 2 \cdot 2 \cdot 2$
2. $4.7 \cdot 4.7 \cdot 4.7$
3. $-9 \cdot -9 \cdot -9 \cdot -9 \cdot -9 \cdot -9$
4. $a \cdot a \cdot a \cdot a \cdot a$
5. $-4 \cdot -4 \cdot -4 \cdot -4 \cdot 6$
6. $7 \cdot 7 \cdot 7 \cdot m \cdot m$
7. $a \cdot b \cdot a \cdot b \cdot a$
8. $5 \cdot 5 \cdot 5 \cdot 5 \cdot 2 \cdot 2 \cdot 2$
9. $-3 \cdot -3 \cdot x \cdot y \cdot x \cdot x \cdot y$

Evaluate each expression.

10. 4^4
11. 5^1
12. $(-3)^4$
13. 6^4
14. 10^3
15. $2^3 \cdot 5^2$
16. $8^3 \cdot (-2)^4$
17. $3^3 \cdot 2^4$

Evaluate each expression if $m = 4$ and $p = 5$.

18. $m^3 + p^3$
19. $p^4 \cdot m$
20. $2 \cdot m^3$
21. p^m

Replace each ■ with <, >, or = to make a true statement.

22. 2^3 ■ 3^2
23. 4^3 ■ 3^4
24. 3^5 ■ 5^3
25. 6^3 ■ 3^6

Problem SOLVING

26. The volume of a cube can be found by multiplying length times width times height. If the length, width, and height are all the same, x inches, write an expression using exponents to find the volume of the cube.

27. A certain type of bacteria splits every hour. How many bacteria will there be after 5 hours?

28. Celia completed the problem below. Describe and correct her error.

$$3^4 = 3 \cdot 4$$
$$= 12$$

Test PREP

29. What is the value of $(0.5)^2$?
    a. 0.1
    b. 0.25
    c. 1.0
    d. 25

30. What is the value of xy^2 if $x = 3$ and $y = 4$?
    a. 12
    b. 24
    c. 36
    d. 48

2.8 Properties of Exponents

Objective: to multiply and divide powers

Exploration Exercise

1. Copy and complete the table.

Exponent Form	Factored Form	Standard Form	Product as a Power
$2^1 \cdot 2^1$	$2 \cdot 2$	4	2^2
$2^2 \cdot 2^3$	$2 \cdot 2 \cdot 2 \cdot 2 \cdot 2$	32	
$3^2 \cdot 3^1$			3^3
$4^3 \cdot 4^3$			

2. What is the relationship between the exponents in the first column and the last column?
3. Write the product of $7^3 \cdot 7^8$ as a single power.

You know that you can write a number in exponential form or factored form. As you saw in the Exploration Exercise above, when you multiply powers with the same base, the exponents are added.

You can write powers with the same base as a power with a single exponent.

$$5^3 \cdot 5^4 = \underbrace{(5 \cdot 5 \cdot 5)}_{3 \text{ factors}} \cdot \underbrace{(5 \cdot 5 \cdot 5 \cdot 5)}_{4 \text{ factors}} = 5^{3+4} = 5^7$$

$$\underbrace{}_{7 \text{ factors}}$$

Multiplying Powers with the Same Base

To multiply powers with the same base, add the exponents.

$$a^m \cdot a^n = a^{m+n} \qquad 2^5 \cdot 2^3 = 2^{5+3} = 2^8$$

Examples
Write each expression as a single exponent.

A. $8^6 \cdot 8^8 = 8^{6+8}$
$ = 8^{14}$

B. $p^7 \cdot p^3 = p^{7+3}$
$ = p^{10}$

You can also divide powers with the same base.

$$\frac{5^6}{5^4} = \frac{\cancel{5} \cdot \cancel{5} \cdot \cancel{5} \cdot \cancel{5} \cdot 5 \cdot 5}{\cancel{5} \cdot \cancel{5} \cdot \cancel{5} \cdot \cancel{5}} = 5 \cdot 5 = 5^{6-4} = 5^2$$

Dividing Powers with the Same Base

To divide powers with the same nonzero base, subtract the exponent of the denominator from the exponent of the numerator.

$$\frac{a^m}{a^n} = a^{m-n} \qquad \frac{3^8}{3^4} = 3^{8-4} = 3^4$$

More Examples
Write each expression using a single exponent.

C. $\dfrac{9^5}{9^2} = 9^{5-2} = 9^3$

D. $\dfrac{r^7}{r^6} = r^{7-6} = r^1 = r$

Try THESE

Tell whether each expression can be simplified to a single exponent.

1. $6^3 \cdot 6^4$
2. $\dfrac{7^3}{5^3}$
3. $8^3 \cdot 3^8$
4. $\dfrac{x^5}{x^2}$

Exercises

Write each expression as a single exponent.

1. $3^6 \cdot 3^4$
2. $m^8 \cdot m^4$
3. $\dfrac{x^9}{x^2}$
4. $v^{11} \cdot v^4$
5. $\dfrac{h^{14}}{h^5}$
6. $4.5^8 \cdot 4.5^2$
7. $\dfrac{9^{21}}{9^9}$
8. $y^5 \cdot y^{14}$
9. $\dfrac{w^{18}}{w^{12}}$
10. $0.5^6 \cdot 0.5^2$
11. $2x^3 \cdot 2x^4$
12. $\dfrac{5m^8}{5m^2}$
13. $x \cdot x^2 \cdot x^2$
14. $6^3 \cdot 6^2 \cdot 6$
15. $a \cdot b^2 \cdot a^4$
16. $m^2n \cdot mn^2$

Replace each ■ with <, >, or = to make a true statement.

17. $2^3 \cdot 2^5$ ■ 2^7
18. $\dfrac{7^4}{7^2}$ ■ 49
19. 5^3 ■ $3^2 \cdot 3^3$
20. $4^4 \cdot 4^2$ ■ 4^8

Problem SOLVING

21. Write 6^{10} as a product of two powers in three different ways.

22. Emma, Erin, and Eden completed the problem to the right. Who completed the problem correctly? Why are the other problems incorrect?

    Emma
    $6^2 \cdot 6^5 = 36^7$

    Erin
    $6^2 \cdot 6^5 = 6^{10}$

    Eden
    $6^2 \cdot 6^5 = 6^7$

2.9 Negative and Zero Exponents

Objective: to use negative and zero exponents

Exploration Exercise
Study the following pattern. What numbers should replace each ■ to continue the pattern?

$$5^3 = 5 \cdot 5 \cdot 5 = 125 \qquad\qquad 5^{-1} = \frac{1}{5^1} = \frac{1}{5} \qquad = \frac{1}{5}$$

$$5^2 = 5 \cdot 5 \qquad\quad = \blacksquare \qquad\qquad 5^{-2} = \frac{1}{5^2} = \frac{1}{5 \cdot 5} \qquad = \frac{1}{\blacksquare}$$

$$5^1 = 5 \qquad\qquad\quad = 5 \qquad\qquad 5^{-3} = \frac{1}{5^3} = \frac{1}{5 \cdot 5 \cdot 5} = \frac{1}{\blacksquare}$$

$$5^0 = 1 \qquad\qquad\quad = 1$$

The Exploration Exercise demonstrates the rules for negative and zero exponents.

> **Negative Exponents**
>
> When a is a nonzero number and n is an integer, $a^{-n} = \frac{1}{a^n}$.
>
> $$a^{-n} = \frac{1}{a^n} \qquad 4^{-3} = \frac{1}{4^3}$$

You can see that this is true by looking at an example.

$$\frac{x^6}{x^9} = \frac{x \cdot x \cdot x \cdot x \cdot x \cdot x}{x \cdot x \cdot x \cdot x \cdot x \cdot x \cdot x \cdot x \cdot x} = \frac{1}{x^3} \text{ or } \frac{x^6}{x^9} = x^{6-9} = x^{-3}$$

So, $\frac{1}{x^3}$ is equal to x^{-3}.

> **Zero Exponents**
>
> For any nonzero number a, $a^0 = 1$.
>
> $$a^0 = 1 \qquad 3^0 = 1$$

$$\frac{x^3}{x^3} = \frac{x \cdot x \cdot x}{x \cdot x \cdot x} = 1 \text{ or } \frac{x^3}{x^3} = x^{3-3} = x^0$$

So, 1 is equal to x^0.

Examples

Simplify each expression.

A. 2^{-4}

$2^{-4} = \dfrac{1}{2^4}$ Write the expression using a positive exponent.

$= \dfrac{1}{16}$ Simplify.

B. $3^{-2} \cdot 4^3$

$3^{-2} = \dfrac{1}{3^2}$ $\qquad$ $4^3 = 4 \cdot 4 \cdot 4$

$= \dfrac{1}{9}$ $\qquad\qquad$ $= 64$

$\dfrac{1}{9} \cdot 64 = \dfrac{64}{9}$

$= 7\dfrac{1}{9}$

C. $5b^0$

$5b^0 = 5 \cdot b^0$ Zero exponent applies to b.

$= 5 \cdot 1$ definition of zero exponent

$= 5$ Multiply.

D. $6x^{-2}$

$6x^{-2} = 6 \cdot x^{-2}$ definition of negative exponent

$= 6 \cdot \dfrac{1}{x^2}$ Exponent applies to x.

$= \dfrac{6}{x^2}$ Multiply.

Try THESE

Fill in each ■ to make a true statement.

1. $3^{■} = \dfrac{1}{3^2}$
2. $5^{■} = 1$
3. $s^{-7} = \dfrac{1}{s^{■}}$
4. $\dfrac{1}{5^9} = 5^{■}$

Exercises

Evaluate each expression.

1. 4^{-2}
2. 8^0
3. 10^{-3}
4. 12^{-1}
5. p^{-8}
6. 2^{-5}
7. $(xy)^0$
8. c^{-9}
9. 5^{-4}
10. $5^2 \cdot 5^{-3}$
11. $4^{-12} \cdot 4^6$
12. $10^8 \cdot 10^{-10}$
13. $3^3 \cdot 2^4$
14. $4^2 \cdot 3^{-3}$
15. $2^3 \cdot 3^{-2}$
16. $5^{-2} \cdot 6^2$

Simplify. Write the expression using only positive exponents.

17. $a^8 \cdot a^{-6}$
18. $c^4 \cdot c^{-10}$
19. $8c^{-6}$
20. $b^{-9} \cdot b^8$

Problem SOLVING

21. Taylor completed the problem to the right. Describe and correct the error in the problem.

$2^{-4} = -2 \cdot -2 \cdot -2 \cdot -2$
$= 16$

2.9 Negative and Zero Exponents   55

2.10 Scientific Notation

Objective: to convert between scientific notation and standard form

Scientists often write very large or very small numbers in **scientific notation**. A number expressed in scientific notation is written as a product. One factor is a number greater than or equal to 1 and less than 10. The other factor is a power of 10.

$64{,}000{,}000 = 6.4000000 \cdot 10{,}000{,}000$ Move the decimal point 7 places to the left and multiply by 10,000,000 or 10^7.
$\phantom{64{,}000{,}000} = 6.4 \cdot 10^7$

The diameter of a grain of sand is as small as 0.0006 meters.

$0.0006 = 0.0006 \cdot 0.0001$ Move the decimal point 4 places to the right and multiply by 0.0001 or 10^{-4}.

$ = 6 \cdot \dfrac{1}{10{,}000}$

$ = 6 \cdot \dfrac{1}{10^4}$

$ = 6 \cdot 10^{-4}$

> The exponent is positive if the decimal point is moved to the left. It is negative if the decimal point is moved to the right.

You can rename a number in scientific notation as a number in standard form as follows.

$6.1 \cdot 10^5 = 6.1 \cdot 100{,}000$

$ = 6.10000$

$ = 610{,}000$ Move the decimal point 5 places to the right.

Examples

A. $3.6 \cdot 10^{-2} = 3.6 \cdot 0.01$
$\phantom{3.6 \cdot 10^{-2}} = 03.6$
$\phantom{3.6 \cdot 10^{-2}} = 0.036$

B. $84{,}320 = 8.4320 \cdot 10{,}000$
$\phantom{84{,}320} = 8.4320 \cdot 10^4$

C. Find the product of $(3.7 \cdot 10^3) \cdot (5.0 \cdot 10^{-1})$.

$= 3.7 \cdot 5.0 \cdot 10^3 \cdot 10^{-1}$ commutative property
$= (3.7 \cdot 5.0) \cdot (10^3 \cdot 10^{-1})$ associative property
$= 18.5 \cdot 10^2$ Add the exponents.
$= 1{,}850$ Simplify.

Try THESE

Tell whether each number is expressed in scientific notation.

1. $3.8 \cdot 10^{-4}$
2. $27.4 \cdot 10^3$
3. $8 \cdot 10^{-2}$
4. $100.1 \cdot 10^2$

Exercises

Write each number in scientific notation.

1. 50,370
2. 68,010,000
3. 8,201,000
4. 131,000
5. 0.0000000714
6. 0.00000916
7. 0.00432
8. 0.000001002

Write each of the following in standard form.

9. $3.25 \cdot 10^2$
10. $4.67 \cdot 10^5$
11. $4.39 \cdot 10^{-3}$
12. $6.8 \cdot 10^{-4}$
13. $5.75 \cdot 10^9$
14. $2.17 \cdot 10^{-8}$
15. $1.18 \cdot 10^{-6}$
16. $9.16 \cdot 10^4$

Compute using scientific notation. Write each answer in standard form.

17. $(6.0 \cdot 10^5) \cdot (3.0 \cdot 10^4)$
18. $(5.1 \cdot 10^{-3}) \cdot (3.2 \cdot 10^3)$
19. $(5 \cdot 10^2) \div (2 \cdot 10^3)$
20. $(7.28 \cdot 10^{-2}) \cdot (9.1 \cdot 10^{-2})$
21. $(2.56 \cdot 10^6) \cdot (3.56 \cdot 10^2)$
22. $(2,965 \cdot 10^7) \div (5.0 \cdot 10^{-3})$
23. $0.0000047 \cdot 0.00006$
24. $0.000027 \div 0.000009$

Problem SOLVING

25. A certain computer can perform 10^5 calculations per second. How many calculations can it perform in 10 seconds?

26. Write a rule for comparing numbers written in scientific notation.

27. The area of Greenland is 840,000 square miles. Write this number in scientific notation.

★28. One nanometer (10^{-9} meters) is about as long as 0.00000003937 inch. Find the approximate number of inches in a meter.

29. In a recent year, the world production of wheat was about $5.2 \cdot 10^8$ metric tons. Canada produced $\frac{1}{25}$ of this. About how much wheat did Canada produce?

30. In a recent year, the population of California was about $2.52 \cdot 10^7$. Its land area is about $4.05 \cdot 10^5$ km². What was the average number of people per square kilometer?

Mind BUILDER

Decimals in Expanded Form

You can write decimals in expanded form using negative exponents.

$634.75 = (6 \cdot 10^2) + (3 \cdot 10^1) + (4 \cdot 10^0) + (7 \cdot 10^{-1}) + (5 \cdot 10^{-2})$

Write each numeral in expanded form using exponents.

1. 147
2. 30.8
3. 11,423.96
4. 0.00256
5. 9.3606

Chapter 2 Review

Language and Concepts

Choose the correct term to complete each sentence.

terminating
repeating
exponent
rational
irrational
base
add
subtract

1. A(n) _____ number can be named by a terminating or repeating decimal.
2. A number used to tell how many times the base is used as a factor is a(n) _____.
3. The decimal $4.\overline{3}$ is an example of a(n) _____ decimal.
4. To multiply powers with the same base, _____ the exponents.
5. The number that is used as a factor in a power is a(n) _____.
6. The decimal 0.125 is an example of a(n) _____ decimal.

Skills and Problem Solving

Rename each fraction or mixed number as a terminating or repeating decimal. Use bar notation for repeating decimals. Section 2.1

7. $\dfrac{3}{8}$
8. $\dfrac{1}{6}$
9. $\dfrac{31}{5}$
10. $7\dfrac{1}{3}$
11. $3\dfrac{2}{9}$

Rename each decimal as a fraction or mixed number in simplest form. Section 2.1

12. 0.4
13. 1.3
14. 4.25
15. 6.32
16. 10.04

Replace each ■ with <, >, or = to make a true statement. Section 2.2

17. 3.5 ■ 3.48
18. -2.8 ■ -2.9
19. $\dfrac{3}{5}$ ■ 0.65
20. 0.34 ■ $\dfrac{1}{3}$

Add or subtract. Write each answer in simplest form. Section 2.3

21. $\dfrac{-1}{9} + \dfrac{-5}{7}$
22. $\dfrac{3}{5} - \dfrac{1}{3}$
23. $\dfrac{11}{12} + \dfrac{-2}{3}$
24. $\dfrac{1}{3} + \dfrac{5}{6}$
25. $5\dfrac{8}{9} - -4$
26. $-6\dfrac{2}{7} + -8\dfrac{3}{14}$
27. $31\dfrac{7}{12} + 15\dfrac{5}{8}$
28. $10\dfrac{5}{18} - 8\dfrac{5}{6}$

Multiply or divide. Write each answer in simplest form. Section 2.4

29. $\dfrac{1}{2} \cdot -1\dfrac{1}{3}$
30. $-10 \cdot 2\dfrac{4}{5}$
31. $\dfrac{5}{8} \div \dfrac{15}{16}$
32. $-5 \div \dfrac{3}{4}$
33. $-2\dfrac{1}{4} \div -7$
34. $3\dfrac{1}{17} \div 1\dfrac{5}{9}$
35. $-5\dfrac{1}{4} \cdot -3\dfrac{2}{3}$
36. $2\dfrac{7}{15} \div -3\dfrac{2}{3}$

Solve each equation. Section 2.5

37. $s + \dfrac{2}{3} = 1\dfrac{1}{2}$
38. $t - 2\dfrac{1}{4} = \dfrac{3}{5}$
39. $4w = 2\dfrac{2}{3}$
40. $\dfrac{x}{3} = 3\dfrac{1}{8}$
41. $x + \dfrac{1}{8} = 2\dfrac{1}{2}$
42. $6x = \dfrac{1}{8}$
43. $\dfrac{-3}{4}x = \dfrac{2}{7}$
44. $x - \dfrac{1}{3} = \dfrac{4}{5}$

Evaluate each expression. Section 2.7

45. 9^3
46. $2^2 \cdot 5^2$
47. $4^3 \cdot 8^2$
48. $3^4 \cdot 3^3$
49. $(-2)^5$
50. $5^2 \cdot (-2)^3$

Write each expression as a single exponent. Section 2.8

51. $x^6 \cdot x^3$
52. $\dfrac{3^5}{3^2}$
53. $4^2 \cdot 4^3 \cdot 4^4$
54. $3m \cdot 3m^3$
55. $0.2^6 \cdot 0.2^8$
56. $\dfrac{x^9}{x^5}$

Evaluate each expression. Section 2.9

57. 5^{-2}
58. $9^0 \cdot 3^{-1}$
59. $3^2 \cdot 5^{-2}$

Write each number in scientific notation. Section 2.10

60. 0.00000000512
61. 605,000,000

Write each of the following in standard form. Section 2.10

62. $5.61 \cdot 10^{-3}$
63. $8.57 \cdot 10^8$

Compute using scientific notation. Write your answer in standard form. Section 2.10

64. $(4 \cdot 10^3) \cdot (6 \cdot 10^6)$
65. $(8.1 \cdot 10^3) \cdot (1.2 \cdot 10^5)$

Solve. Sections 2.3, 2.6

66. Four employees can make twenty-four pizzas in 1 hour. How many pizzas can eight employees make in 20 minutes?

67. Carol is baking cookies, and she needs $2\frac{1}{4}$ cups of flour, $\frac{2}{3}$ cup of sugar, $\frac{1}{4}$ cup of brown sugar, and $\frac{1}{16}$ cup of baking soda. What is the total amount, in cups, of the dry ingredients?

68. If $\frac{1}{3}$ of the Petunias are red and $\frac{1}{2}$ are purple, what fraction of the Petunias are other colors?

69. You decide to number the 42 pages in your journal from 1 to 42. How many digits do you have to write?

Chapter 2 Test

Write each fraction or mixed number as a terminating or repeating decimal. Use bar notation for repeating decimals.

1. $\dfrac{15}{12}$
2. $\dfrac{17}{40}$
3. $\dfrac{17}{3}$
4. $6\dfrac{4}{15}$

Write each decimal as a fraction or mixed number in simplest form.

5. 0.36
6. 2.18
7. 5.410
8. 14.095

Order the numbers in each list from least to greatest.

9. $\dfrac{2}{3}$, 0.775, 0.6, $\dfrac{5}{8}$
10. 4.6, $\dfrac{30}{4}$, 4.75, $\dfrac{28}{6}$
11. $\dfrac{-96}{10}$, -9.5, $-9\dfrac{2}{3}$, $\dfrac{-28}{3}$

Compute. Write each answer in simplest form.

12. $\dfrac{2}{5} + \dfrac{3}{10}$
13. $\dfrac{-4}{5} + -3\dfrac{1}{6}$
14. $\dfrac{2}{3} - \dfrac{-3}{7}$
15. $\dfrac{2}{3} \cdot \dfrac{-4}{5}$
16. $\dfrac{-1}{2} + \dfrac{-5}{6}$
17. $\dfrac{4}{9} \div \dfrac{2}{3}$
18. $9\dfrac{2}{3} + 8\dfrac{1}{16}$
19. $9 - 6\dfrac{4}{7}$
20. $14\dfrac{7}{16} + \dfrac{3}{4}$
21. $\dfrac{-5}{7} \cdot -27$
22. $\dfrac{6}{7} \cdot 3\dfrac{1}{9}$
23. $2\dfrac{4}{5} \div -7$
24. $\dfrac{-3}{4} + \dfrac{-5}{6}$
25. $\dfrac{3}{16} \cdot \dfrac{8}{9}$
26. $19\dfrac{1}{6} - 13\dfrac{7}{8}$
27. $-2\dfrac{4}{9} \div -1\dfrac{5}{6}$

Solve each equation.

28. $w + \dfrac{5}{8} = 2\dfrac{3}{4}$
29. $x \div 7 = -2\dfrac{1}{3}$
30. $7\dfrac{2}{7} \cdot s = 12\dfrac{3}{4}$
31. $x - 4\dfrac{3}{5} = 8$

Evaluate each expression.

32. $4^{-3} \cdot 8^2$
33. $\dfrac{9^2}{7^0}$
34. $5^3 \cdot 5^2$

Solve.

35. Ted Andrews bought 15 shares of Clark Company stock at $14\dfrac{7}{8}$. He sold them a month later at $20\dfrac{3}{8}$. What was his profit or loss? Show all work, and explain your answer.

36. How many rectangles are there in the figure below?

37. In a tennis tournament, each athlete plays one match against each of the other athletes. There are 12 athletes scheduled to play in the tournament. How many matches will be played?

Change of Pace

History

The Greek mathematician Euclid wrote about numbers and divisibility in his book *The Elements*. He gave a method for finding the GCF of groups of numbers. This method is now called the **Euclidean Algorithm**. This algorithm works as follows.

Find the GCF of 30 and 145.

Step 1	Step 2	Step 3
Divide the greater number by the lesser number.	Divide the lesser number by the remainder.	Continue dividing the divisor by the remainder until you get a remainder of 0. The last nonzero remainder is the GCF.
$\begin{array}{r}4\\30\overline{)145}\\-120\\\hline 25\end{array}$	$\begin{array}{r}1\\25\overline{)30}\\-25\\\hline 5\end{array}$ last nonzero remainder	$\begin{array}{r}5\\5\overline{)25}\\-25\\\hline 0\end{array}$ remainder of 0

The GCF of 30 and 145 is 5, the last nonzero remainder. Check this using prime factorization.

You can use the GCF to find the least common multiple (LCM). The LCM of 30 and 145 is 870. Check this using prime factorization.

$$\text{LCM of 30 and 145} = \frac{30 \times 145}{\text{GCF of 30 and 145}} = \frac{4{,}350}{5} \text{ or } 870$$

Use the Euclidean Algorithm to find the GCF of each pair of numbers. Then use the GCF to find the LCM of each pair of numbers.

1. 9, 75
2. 16, 108
3. 81, 315
4. 110, 63
5. 456, 759
6. 136, 232
7. 567, 432
8. 975, 364

9. Which pair of numbers in problems 1–8 above are relatively prime (GCF of 1)?

Find the GCF of each group of numbers. Use the Euclidean Algorithm with any two numbers. Then apply the algorithm again with the third number and the GCF of the first two.

10. 72, 90, 96
11. 312; 468; 1,012
12. 714; 2,030; 2,205

Cumulative Test

1. Which number replaces the ■ to make a true sentence?
   $3 \cdot \blacksquare = (3 \cdot 6) + (3 \cdot 4)$
   a. 4
   b. 6
   c. 10
   d. 30

2. Which of the following is $\frac{12}{20}$ in simplest form?
   a. $\frac{1}{2}$
   b. $\frac{3}{5}$
   c. $\frac{6}{10}$
   d. $1\frac{2}{3}$

3. Which of the following is 6^3 written as a product?
   a. $3 \cdot 6$
   b. $6 \cdot 3$
   c. $6 \cdot 6 \cdot 6$
   d. $6 + 6 + 6$

4. Solve.
   $3 + \frac{4}{m} = 5$
   a. 0.5
   b. 1
   c. 2
   d. 8

5. Which of the following is divisible by 4?
   a. 98
   b. 136
   c. 149
   d. none of the above

6. Which statement is true?
   a. $\frac{2}{3} < \frac{3}{4}$
   b. $\frac{1}{8} < \frac{2}{16}$
   c. $\frac{4}{5} > \frac{16}{20}$
   d. $\frac{5}{6} > \frac{25}{30}$

7. The number 50 is divided by a number between 1 and 2. Which of the following best describes the quotient?
   a. between 25 and 50
   b. between 50 and 100
   c. greater than 100
   d. less than 25

8. The students attending a school can be divided into equal teams of 12 or 15. Which of the following is the least number of students who could attend the school?
   a. 27
   b. 60
   c. 180
   d. 600

9. The formula $W = 0.8 \cdot (220 - A)$ can be used to find a working heart rate for a person of age A. Approximately what should the working heart rate be for a 34-year-old person?
   a. 140
   b. 150
   c. 176
   d. 186

10. How much more would Bill pay to rent a 10-speed bicycle for 3 hours than to rent a 3-speed bicycle for the same time?
    a. $0.50
    b. $1.50
    c. $8.00
    d. $3.50

Smith's Bicycle Rental		
bicycle	per hour	per day
1-speed	$2.50	$11
3-speed	$3.00	$13
10-speed	$3.50	$14

CHAPTER 3

Real Numbers

Julia Marshall
Kentucky

3.1 Squares and Square Roots

Objective: to find the square and square roots of numbers

The region the square map represents has 81 square miles. What is the length of each side of the region?

A number, such as 81, is a **perfect square** because it is the square of a whole number. The **square root** of a number is one of two equal factors of a number.

What number multiplied by itself gives 81? Since 81 = 9 • 9, the length of each side of the region is 9 miles.

Since -9 • -9 = 81, another square root of 81 is -9. Notice that a negative sign is used to indicate the negative square root.

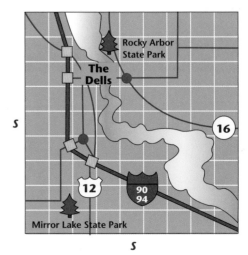

$\sqrt{81} = 9$ The positive square root of 81 is 9.
$-\sqrt{81} = -9$ The negative square root of 81 is -9.

Examples

A. Find the two square roots of 25.

5 and -5

Not all numbers are perfect squares. To find the square root of a number that is not a perfect square, you can estimate to the nearest integer.

B. Estimate the square root of 34 to the nearest integer.

Since $\sqrt{34}$ is closer to $\sqrt{36}$, $\sqrt{34} \approx 6$.

Try THESE

Find the square of each number.

1. 2   2. 5   3. 14   4. 25   5. 81   6. 144

Exercises

Find the two square roots of each number.

1. 64
2. 49
3. 400
4. $\dfrac{1}{100}$
5. $\dfrac{1}{4}$
6. 256
7. $\dfrac{1}{144}$
8. $\dfrac{9}{25}$
9. 625
★10. 1.21
★11. 6.25
★12. 1.69

Estimate to the nearest integer.

13. $\sqrt{8}$
14. $-\sqrt{15}$
15. $\sqrt{42}$
16. $\sqrt{115}$
17. $-\sqrt{45}$
18. $\sqrt{78}$
19. $-\sqrt{93}$
20. $\sqrt{56}$

Problem SOLVING

21. Rosa makes a square rug that has an area of 400 square inches. What is the length of each side of the rug?

22. Wei-Min makes a square quilt that has an area of 100 square feet. It has 25 square blocks, each the same size. What is the length of each side of a block?

Constructed RESPONSE

★23. The difference between two perfect squares is 45. The difference between their square roots is 3. What are the squares? What are their square roots? Show all work, and explain your reasoning.

24. The difference between two perfect squares is 95. The difference between their square roots is 5. What are the squares? What are their square roots? Show all work, and explain your reasoning.

Test PREP

You can use the formula $d = \sqrt{1.5h}$ to estimate the distance d, in miles, to a horizon line when your eyes are h feet above the ground.

25. Brock is sitting in a lifeguard chair 11 feet above the beach. About how far can Brock see to the horizon line?

    a. 14 feet
    b. 4 feet
    c. 11 feet
    d. 8 feet

3.1 Squares and Square Roots

3.2 Real Numbers

Objective: to identify rational and irrational numbers

You know about integers and rational numbers. **Integers** are numbers, such as 3, 0, -5, and 91. **Rational numbers** are numbers that can be named as the quotient of two integers. The following are rational numbers.

$$5 \qquad -0.42 \qquad \frac{2}{5} \qquad 3.6$$

Rational numbers whose square roots are rational numbers are called perfect squares. For example, since 9 • 9 = 81, one square root of 81 is 9. So, 81 is a perfect square. Another square root of 81 is -9, which is also a rational number.

What about the square roots of rational numbers that are not perfect squares? Consider $\sqrt{2}$. Can you find a rational number that when squared equals 2?

$$(1.5)^2 = 2.25 \qquad (1.4)^2 = 1.96 \qquad (1.42)^2 = 2.0164$$

It appears that the square root of 2 is not a rational number.

$$\sqrt{2} = 1.4142136 \ldots$$

▶ This decimal continues forever without any pattern of repeating digits.

A number that can be named only by a nonterminating, nonrepeating decimal is called an **irrational number**. Irrational numbers together with rational numbers form the set of **real numbers**. The web below shows the relationship between the numbers in the real number system.

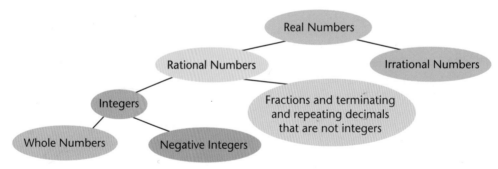

Examples

	Number	Integer	Rational	Irrational	Real
A.	-7	✓	✓		✓
B.	0.12121212...		✓		✓
C.	$\sqrt{14}$			✓	✓
D.	0.1011011101111...			✓	✓
E.	$\sqrt{9}$	✓	✓		✓
F.	$-\frac{4}{3}$		✓		✓

Try THESE

Identify the number that is not the same as the others in each group.

1. a. 5.85   b. 63.4   c. 8.52624...   d. 27.5
2. a. $\sqrt{25}$ b. $\sqrt{41}$ c. $\sqrt{49}$ d. $\sqrt{81}$

Exercises

Name the sets of numbers to which each number belongs—*integers, rationals, irrationals,* or *reals.*

1. -6
2. $\frac{1}{2}$
3. $\frac{9}{3}$
4. 15
5. -0.3333...
6. $\sqrt{12}$
7. $\sqrt{25}$
8. 0.125
9. 0.36945
10. $0.\overline{53}$
11. 0.6125
12. 3.1416...
13. $\sqrt{\frac{100}{36}}$
14. $-\sqrt{16}$
15. $-\sqrt{43}$
16. 1.123124125
17. $0.\overline{571428}$
18. $2.9\overline{36}$
19. 5.45445444...

Write whether each sentence is *true* or *false.*

20. Every integer is a real number.
21. The number 0 is a rational number.
22. Every real number is a rational number.
23. Some irrational numbers are also rational numbers.
24. Every rational number has a square root.
25. Every irrational number is a real number.
26. Every integer is a rational number.

Problem SOLVING

★27. Mike's living room is a square. It has an area of 144 square feet. How many square yards of carpeting does it take to carpet the room?

28. Jane makes a square rug that has an area of 900 square inches. What is the length of each side of the rug?

MiXeD REVIEW

Compute.

29. $\frac{3}{5} - \frac{1}{4}$
30. $3\frac{5}{8} + 2\frac{7}{12}$
31. $\frac{3}{4} \div \frac{1}{3}$
32. $\frac{4}{7} \cdot \frac{3}{5}$
33. $10 \cdot \frac{1}{4}$
34. $\frac{3}{7} \div \frac{12}{21}$
35. $5\frac{3}{4} - 2\frac{1}{8}$
36. $\frac{5}{6} + \frac{7}{9}$

3.3 The Pythagorean Theorem

Objective: to use the Pythagorean Theorem

Mike Moore helped survey the three subdivisions in the town of Royalton that bound Crockett Park. Since Austin Avenue and Bowie Street meet at a right angle, △ABC is a right triangle. Study the drawing below, and find the area of each subdivision to discover an important property of right triangles.

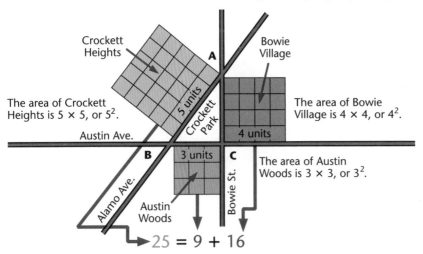

Add the areas of Austin Woods and Bowie Village. Does the sum equal the area of Crockett Heights? This is an example of a property of right triangles called the **Pythagorean theorem**.

 Pythagorean Theorem
In a right triangle, the square of the length of the hypotenuse is equal to the sum of the squares of the lengths of the other two sides.

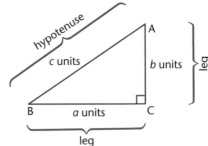

Example

You can use the Pythagorean theorem to find the length of the missing side of the right triangle.

$$c^2 = a^2 + b^2$$
$$13^2 = a^2 + 12^2$$
$$169 = a^2 + 144$$
$$169 - 144 = a^2 + 144 - 144$$
$$25 = a^2$$
$$\sqrt{25} = a$$
$$5 = a$$

Side a is 5 meters long.

68   3.3 The Pythagorean Theorem

Try THESE

Use the Pythagorean theorem to write an equation for each triangle. If necessary, round answers to the nearest hundredth.

1.
2.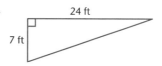

Exercises

Use the Pythagorean theorem to find the missing length of each right triangle. Round to the nearest hundredth, if necessary.

1. $a = 3$ in., $b = 4$ in.
2. $a = 8$ m, $b = 6$ m
3. $a = 40$ in., $c = 41$ in.
4. $a = 12$ cm, $b = 9$ cm
5. $b = 80$ m, $c = 89$ m
★6. $a = 15$ in., $b = 3$ ft
7. $a = 15$ ft, $c = 17$ ft
8. $b = 7$ m, $c = 10$ m
9. $a = 21$ yd, $b = 28$ yd

Problem SOLVING

Determine whether each triangle with sides of given lengths is a right triangle.

10. 7 m, 15 m, 18 m
11. 21 in., 28 in., 35 in.
12. 20 ft, 24 ft, 36 ft

13. Which rectangle has the longer diagonal?

 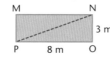

★14. The hypotenuse of a right triangle is 15 cm, and one of its legs is 6 cm. Find the length of the other leg.

Mid-Chapter REVIEW

Determine whether each number is *rational* or *irrational*.

1. 14.1
2. $\frac{36}{121}$
3. $0.\overline{54}$
4. $\sqrt{19}$
5. $-\sqrt{49}$

6. Name the hypotenuse and legs.

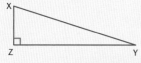

7. Find the length of the red side.

8. How long is the brace on the gate?

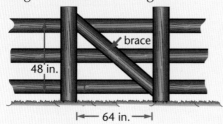

9. How long is the longest side of the sail?

3.3 The Pythagorean Theorem

3.4 Using the Pythagorean Theorem

Objective: to solve problems using the Pythagorean theorem

Russell Gossman is training for the Olympics. He wants to know the distance across Lake Royal, so he can write in his training logbook how far he rows each day. First, he finds the lengths of $\overline{AB}$ and $\overline{AC}$. Then, he uses the Pythagorean theorem to find the length of $\overline{BC}$.

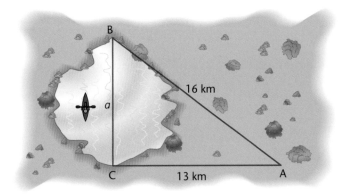

$$c^2 = a^2 + b^2$$
$$16^2 = a^2 + 13^2$$
$$256 = a^2 + 169$$
$$256 - 169 = a^2 + 169 - 169 \quad \text{Subtract 169 from both sides.}$$
$$87 = a^2$$
$$\sqrt{87} = a$$
$$9.327 = a$$

The distance across the lake is about 9.33 km.

More Examples

A. How wide is the television screen?

$$c^2 = a^2 + b^2$$
$$12^2 = 10^2 + b^2$$
$$144 = 100 + b^2$$
$$144 - 100 = 100 - 100 + b^2$$
$$44 = b^2$$
$$\sqrt{44} = b$$
$$6.633 = b$$

The television screen is about 6.633 in. wide.

B. How far above the ground is the kite?

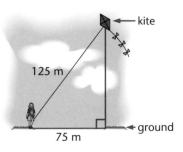

$$c^2 = a^2 + b^2$$
$$125^2 = a^2 + 75^2$$
$$15{,}625 = a^2 + 5{,}625$$
$$15{,}625 - 5{,}625 = a^2 + 5{,}625 - 5{,}625$$
$$10{,}000 = a^2$$
$$\sqrt{10{,}000} = a$$
$$100 = a$$

The kite is 100 m above the ground.

Try THESE

Write an equation that can be used to answer each question. Then solve.

1. How far is the weather balloon from the weather station?

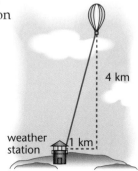

2. How far above the ground does the ladder touch the house?

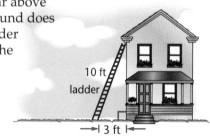

Exercises

Round your answers to the hundredths place.

1. What is the distance between the feet of the sawhorse?

2. How long is each rafter?

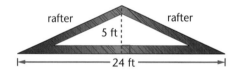

Problem SOLVING

Solve. Round your answers to the nearest hundredth.

3. Anita hikes 7 miles due east and then 3 miles due north. What is the shortest distance back to where she began?

4. The foot of a ladder is placed 5 feet from a building. The ladder is 13 feet long. How high does the ladder reach on the building?

5. A baseball diamond is a square. The distance from home plate to first base is 90 feet. Find the distance from home plate to second base.

6. Wire is stretched from the top of a 26-foot pole to a point on the ground that is 15 feet from the bottom of the pole. How long is the wire?

7. A clock has a minute hand that is 1.5 meters long and an hour hand that is 1.2 meters long. What is the distance between the ends of the hands at 9 o'clock?

★8. Mel takes a shortcut to school by walking diagonally across an empty lot. The rectangular lot is 20 meters wide and 40 meters long. How much shorter is the shortcut than a route on the sides of the lot? Explain.

3.4 Using the Pythagorean Theorem

Problem Solving

What Color is It?

You are shown three views of the same eight-sided diamond. All the faces are equilateral triangles. What color is the side marked with a question mark?

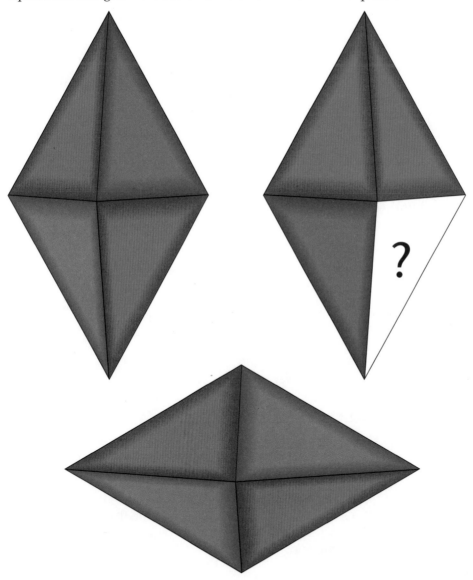

Extension

Make up a puzzle similar to this one.

Cumulative Review

Estimate.

1. 7,846 + 3,782
2. 4,180 − 1,347
3. 278 × 44
4. 278)11,964
5. 4.1)3.55

Compute.

6. $49.03 − 17.26
7. 47.5 + 8.74 + 0.198
8. 27.2 − 8.88

9. 42 × 8.4
10. 0.47 × 0.02
11. 33.78 × 5.9
12. 67,900 × 0.24

13. 0.6)5.94
14. 12)27
15. 4.6)9.2
16. 0.72)4.68
17. -9 + -7
18. -9 + 5.1
19. -2 − -7
20. 6 − 12
21. -4 • -9.5
22. -2 • 7
23. -16 ÷ 2
24. -72 ÷ -1.8

Find the greatest common factor (GCF) of each group of numbers.

25. 20, 36
26. 32, 45
27. 45, 60
28. 12, 18, 24

Find the least common multiple (LCM) of each group of numbers.

29. 6, 8
30. 8, 18
31. 5, 15
32. 2, 3, 10

Compute. Write each answer in simplest form.

33. $\frac{7}{9} + \frac{1}{4}$
34. $5\frac{1}{2} + 2\frac{5}{8}$
35. $\frac{4}{5} - \frac{1}{6}$
36. $2\frac{1}{7} - 1\frac{2}{3}$
37. $\frac{5}{8} \cdot \frac{4}{5}$
38. $2\frac{2}{3} \cdot 3\frac{3}{4}$
39. $\frac{3}{8} \div \frac{3}{4}$
40. $5\frac{1}{2} \div 2$

Solve each equation. Check your solution.

41. $a - 8 = 25.5$
42. $17 + w = 65$
43. $5.9 = h - 9.6$
44. $9r = 10$
45. $\frac{t}{7} = 6.05$
46. $2s + 5 = 13$

Solve.

47. A dressmaker cuts a piece of ribbon $8\frac{5}{8}$ inches long from a piece $12\frac{1}{2}$ inches long. How many inches of ribbon are left?

48. Doug spent $2\frac{1}{5}$ hours on his homework. Emma spent $3\frac{3}{4}$ hours on her homework. How much time did they spend altogether?

49. A box of detergent contains 20 cups. How many loads of laundry can be washed if each load uses $\frac{3}{4}$ of a cup of detergent?

50. It takes Teresita 12 minutes to saw a log into three pieces. How long would it take her to saw a log into four pieces?

3.5 30°–60° Right Triangles

Objective: to find the length of a side of a 30°–60° right triangle

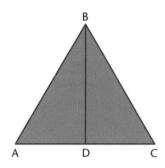

You know the following angle measures.

m∠A = 60°   △ABC is an equilateral triangle.

m∠ABD = 30°   ∠ABC is bisected to form ∠ABD and ∠CBD.

m∠ADB = 90°   m∠ADB + m∠CDB is 180°, and these angles are congruent (corresponding parts of congruent triangles).

So, △ABD is a right triangle. More specifically, it is called a **30°–60° right triangle**. Study the following to learn an important property of such triangles.

$\overline{AD}$ and $\overline{DC}$ are corresponding parts of congruent triangles.
So, $\overline{AD} \cong \overline{DC}$, or the length of $\overline{AD}$ is one-half the length of $\overline{AC}$.
Since $\overline{AC} \cong \overline{AB}$, the length of $\overline{AD}$ (the side opposite the 30° angle of △ABD) is one-half the length of $\overline{AB}$ (the hypotenuse).

 Rule In any 30°–60° right triangle, the length of the side opposite the 30° angle is one-half the length of the hypotenuse.

More Examples

A. Find the length of the side opposite the 30° angle in the triangle below.

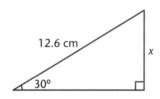

$x = 0.5 \cdot 12.6$

$x = 6.3$

The side opposite the 30° angle is 6.3 cm long.

B. Find the length of the hypotenuse in the triangle below.

$2 = 0.5 \cdot h$

$\dfrac{2}{0.5} = \dfrac{0.5h}{0.5}$

$4 = h$

The hypotenuse is 4 mm long.

Try THESE

The length of the hypotenuse of a 30°–60° right triangle is given. Find the length of the side opposite the 30° angle.

1. 8 m
2. 6.7 cm
3. 4.59 mm
4. 4.5 mi
5. $6\frac{5}{8}$ in.
6. $17\frac{1}{2}$ in.
7. $3\frac{1}{3}$ yd
8. 4.38 in.

Exercises

The length of the side opposite the 30° angle of a right triangle is given. Find the length of the hypotenuse.

1. 9 m
2. 6.9 in.
3. 4.27 km
4. $4\frac{1}{2}$ in.
5. $5\frac{3}{8}$ in.
6. 12 m
7. 4.39 m
8. 8.5 cm

Find the length of the red side of each triangle.

9.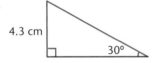
10.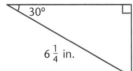
11.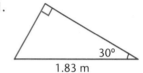
12.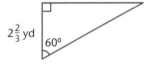
13.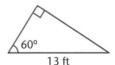
14.

Problem SOLVING

15. How long is each rafter?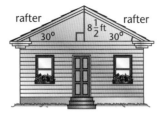

★16. How tall is the water tower?

17. Mrs. Sisk makes a gate for her picket fence. It is 1 meter wide and 1.5 meters high. How long is the diagonal brace for the gate?

★18. Use the Pythagorean theorem to write a rule for finding the length of the side opposite the 60° angle of a 30°–60° right triangle.

★19. The sum of the squares of two integers is 289. What are the integers?

3.5 30°–60° Right Triangles

3.6 45°–45° Right Triangles

Objective: to find the length of a side of a 45°–45° right triangle

A throw from home plate to second base bisects the right angles formed by the base lines. The two triangles formed are congruent isosceles right triangles.

You know the following angle measures.

$m\angle B = 90°$ ABCD is a square.

$m\angle BAC = m\angle BCA$ $\angle A$ and $\angle C$ are bisected to
$= 45°$ form $\angle BAC$ and $\angle BCA$.

So, $\triangle ABC$ is a **45°–45° right triangle**.

Study the following to learn an important property of such triangles.

Let s = the measure of each side of a square.

Let d = the measure of a diagonal of the square.

$d^2 = s^2 + s^2$ Pythagorean theorem

$d^2 = 2s^2$

$d = \sqrt{2s^2}$

$d = s\sqrt{2}$

 Rule   In any 45°–45° right triangle, the length of the hypotenuse is $\sqrt{2}$ times the length of the side opposite the 45° angle.

Another Example

The distance from home plate to second base on a softball diamond is $60\sqrt{2}$ ft. How far is it between bases?

$\triangle PTH$ is a right isosceles triangle.
$m\angle 1 = m\angle 2 = 45°$

Let $PT = d$.

$d = s\sqrt{2}$

$60\sqrt{2} = s\sqrt{2}$

$\dfrac{60\sqrt{2}}{\sqrt{2}} = \dfrac{s\sqrt{2}}{\sqrt{2}}$

$60 = s$

The distance between bases is 60 ft.

Try THESE

The length of a side opposite the 45° angle of a right triangle is given. Find the length of the hypotenuse.

1. 1 ft
2. 3 cm
3. 2 in.
4. 31.2 m
5. $4\frac{2}{3}$ yd

Exercises

The length of the hypotenuse of a 45°–45° right triangle is given. Find the length of the other sides.

1. $11\sqrt{2}$ yd
2. $12\sqrt{2}$ in.
3. $5.2\sqrt{2}$ cm
4. $\frac{\sqrt{2}}{2}$ m
5. $\frac{3\sqrt{2}}{2}$ ft

Find the value of x.

6.
7.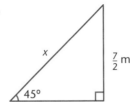
8.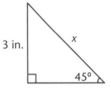

Problem SOLVING

9. A floor is covered with square tiles arranged in a diamond pattern. The distance between opposite corners of a tile is $50\sqrt{2}$ cm. What is the area of the tile?

★10. The sum of the squares of the lengths of all sides of a rectangle is 1,458 units. Find the length of a diagonal of the rectangle.

Mixed REVIEW

Change each fraction or mixed number to a decimal.

11. $\frac{1}{2}$
12. $\frac{4}{5}$
13. $\frac{5}{9}$
14. $\frac{13}{10}$
15. $\frac{5}{6}$
16. $\frac{19}{20}$
17. $2\frac{5}{8}$
18. $3\frac{21}{25}$
19. $10\frac{3}{4}$
20. $5\frac{1}{12}$
21. $4\frac{9}{11}$
22. $6\frac{4}{15}$

Mind BUILDER

Pythagorean Triples

Three numbers that satisfy the Pythagorean theorem are called **Pythagorean triples**. The set of numbers 5, 12, and 13 is a Pythagorean triple because $5^2 + 12^2 = 13^2$.

State whether each set of numbers is a Pythagorean triple. Write *yes* or *no*.

1. 3, 4, 5
2. 8, 10, 13
3. 45, 60, 75
★4. 100, 240, 260

5. Find a Pythagorean triple with one leg 7 units long and the other two sides between 20 and 30 units long.

3.7 Problem-Solving Strategy: Using Logical Reasoning

Objective: to use logical reasoning to problem solve

Mark, Tanya, Scott, and Peggy each have a different favorite sport. Mark's favorite is not track. Tanya does not like volleyball or track. Someone likes basketball. Peggy dislikes football. Scott likes the sport that Peggy dislikes. What is the favorite sport of each person?

You know what sport one person likes and what sport the other three people dislike. You need to find each person's favorite sport.

To find each person's favorite sport you need to organize all the information. Use a table and logical reasoning.

	Track	Volleyball	Basketball	Football
Mark				No
Tanya				No
Scott	No	No	No	Yes
Peggy				No

Since Scott's favorite sport is football, the other three sports are not his favorite. Football cannot be anyone else's favorite.

Continue using the clues to fill in the table.

	Track	Volleyball	Basketball	Football	
Mark	No	Yes	No	No	Step 2
Tanya	No	No	Yes	No	Step 3
Scott	No	No	No	Yes	Step 4
Peggy	Yes	No	No	No	

The solution is: Mark—volleyball, Tanya—basketball, Scott—football, and Peggy—track.

The solution is reasonable, since there is only one *yes* in every row and column.

Try THESE

Solve.

1. A square, a rhombus, a trapezoid, and a parallelogram are blue, red, yellow, and green but not necessarily in that order. The green figure has four equal angles. The trapezoid is not red. The color of the parallelogram has more than four letters. Match each shape to its color.

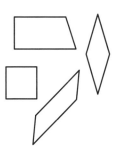

Solve

1. Six good friends live in apartments **A, B, C, D, E,** and **F**. Julie's apartment is just after Len's. Robert lives between Ed and Betty. Sue lives in one of the apartments. Robert lives in **C**. Len's apartment is not **A, B, C,** or **D**. Ed lives two doors before Betty. Who lives in which apartment?

★2. Bev, Andy, and Peta are a nurse, an electrician, and a computer programmer. Bev does not like illness. Andy is not 24 years old, but his friend is. The electrician is not 28 years old. Bev is 28 years old, and her friend, Andy, is the one who works with sick people. What is the career of each person?

★3. Write a logic problem similar to the ones above.

Use the diagram to solve problems 4–5.

4. If $\overline{BC}$ is the shortest side of $\triangle ABC$, what can you conclude about the measure of $\angle CAB$?

5. Name the hypotenuse of $\triangle ABC$ in two different ways.

6. *True* or *false?* The measure of a right angle is always 90°.

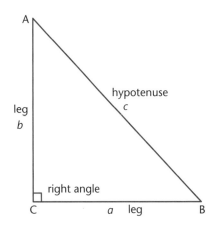

Mind BUILDER

Puzzle

Each whole number from 1 through 15 can be expressed using the numerals 1, 2, 3, and 4 exactly once and any of the operation signs. For example, $1 = (1 + 4) \div (2 + 3)$. Write an expression for each of the others.

Chapter 3 Review

Language and Concepts

Choose a term or number from the list at the right to correctly complete each sentence.

1. A(n) _____ number can be named by a nonterminating, nonrepeating decimal.
2. The hypotenuse is the side opposite the _____ angle of a right triangle.
3. Either of the two sides of a right triangle is called a(n) _____.
4. In any 30°–60° right triangle, the length of the side opposite the _____ angle is one-half the length of the hypotenuse.
5. A(n) _____ number is either a rational number or an irrational number.
6. In any 45°–45° right triangle, the length of the hypotenuse is _____ times the length of the side opposite the 45° angle.
7. In the general form of the Pythagorean theorem $a^2 + b^2 = c^2$, c represents the _____.

$\frac{1}{2}$
$\sqrt{2}$
$\sqrt{3}$
30°
45°
60°
90°
hypotenuse
irrational
leg
rational
real

Skills and Problem Solving

Find the two square roots of each number. Section 3.1

8. 144
9. 196
10. $\frac{1}{9}$
11. $\frac{4}{49}$

State the sets of numbers to which each number belongs—*integers, rationals, irrationals,* or *reals.* Section 3.2

12. 0.20220222...
13. $\sqrt{121}$
14. $\frac{1}{4}$
15. -79
16. $-\sqrt{32}$

Use the Pythagorean theorem to find the missing length for each right triangle. Round answers to the nearest tenth. Section 3.3

17. $a = 20$ cm, $b = 21$ cm
18. $a = 15$ in., $c = 39$ in.
19. $a = 2$ m, $b = 5$ m
20. $a = 6$ yd, $b = 6$ yd
21. $b = 16$ ft, $c = 34$ ft
22. $a = 7$ ft, $b = 7$ ft
23. $b = 10$ ft, $c = 26$ ft
24. $a = 4$ km, $b = 7$ km
25. $a = 16$ cm, $b = 65$ cm
26. $a = 9$ cm, $b = 3$ cm
27. $a = 5$ in., $b = 5$ in.
28. $b = 14$ ft, $c = 18$ ft

**Find the length of the red side of each of the following triangles.
Sections 3.3, 3.5–3.6**

29.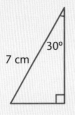

30.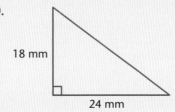

31.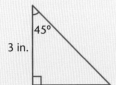

32.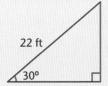

33.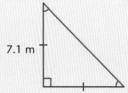

Solve. Section 3.4

34. April is standing 28 feet from a point directly below her kite. The string attached to the kite is 53 feet long. How high is the kite flying?

35. An airplane travels due east 65 miles and then due north 72 miles. How far is the airplane from its starting point?

Complete a table to solve the problem. Section 3.7

36. Jeff, Billy, and Martin are the boyfriends of three sisters, Lise, Karen, and Alicia. Karen is a blond. Martin does not like girls with blond hair. Alicia only sees her boyfriend on weekends. Jeff, Alicia, and Martin go to the same college. Who dates Karen?

Chapter 3 Test

Name all sets of numbers to which each number belongs.

1. $\frac{4}{3}$
2. $\sqrt{18}$
3. $0.10010001\ldots$
4. $-\sqrt{25}$
5. 3.4141

Find both square roots of each number.

6. 256
7. $\frac{1}{100}$
8. $\frac{36}{49}$

Find the length of the missing side in each of the following triangles.

9.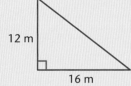
10.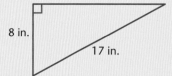
11.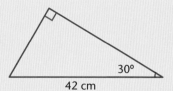

Find the length of the red side in each of the following triangles.

12.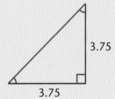
13.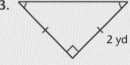
14.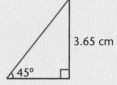

Solve.

15. What is the missing length of the side of the sail?

    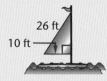

16. What is the height of the triangle?

    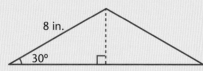

17. Justin, Yuji, Tom, and Ken love sports—soccer, baseball, tennis, and football. Justin does not like shoulder pads. Yuji and Tom prefer kicking to hitting. Ken and Tom do not like to wear shorts. Match each person with the sport he plays.

Change of Pace

Logic

Logic deals with the formal principles of reasoning. Most discussions that use logic begin with a **statement**. Consider the following statements.

1. Columbus is the capital of Ohio.
2. Columbus is not the capital of Ohio.
3. He weighs 140 pounds.
4. $10 \div 5 = 2$

Statements may be true, such as numbers 1 and 4 above, or false, such as number 2. Some statements, like number 3, are neither true nor false. The truth or falsity of a statement is called its **truth value**.

    Charlie is a horse. Charlie has four legs.   True

    Charlie has four legs. Charlie is a horse.   No conclusion

We can form a new statement called a **conditional**, by connecting two statements with the words *if* and *then*. "If Charlie is a horse, then Charlie has four legs." is a conditional. New conditions can be formed as follows.

    If Charlie is not a horse, then Charlie does not have four legs.   No conclusion

    If Charlie does not have four legs, then Charlie is not a horse.   True

State the truth value for each of the following.

1. Washington, DC, is the capital of the United States.
2. The Mississippi River flows through the state of Maine.
3. $17 \div 5 = p$
4. $\frac{1}{2} \cdot \frac{1}{4} = \frac{1}{6}$
5. The number 21 is prime.
6. Mars has two Moons.

Determine the truth value for each of the following. Use the original true statement "If Ruff is a beagle, then Ruff is a dog."

7. Ruff is a dog. Ruff is a beagle.
8. Ruff is a beagle. Ruff is a dog.
9. If Ruff is a dog, then Ruff is a beagle.
10. If Ruff is not a dog, then Ruff is not a beagle.

Cumulative Test

1. Which number has a square root of 9?
   a. -81
   b. 3
   c. 18
   d. 81

2. A board 3 m long is bisected. How long is each new piece?
   a. 1.5 m
   b. 3 m
   c. 6 m
   d. none of the above

3. What is the value of $(-2 \cdot 7) + 6 \cdot 2$?
   a. 26
   b. -2
   c. -16
   d. -26

4. Sam bought some stamps. He gave $\frac{1}{2}$ of the stamps to his brother and used $\frac{1}{3}$ of the remaining stamps. He now has 12 stamps. How many did he buy?
   a. 24
   b. 36
   c. 40
   d. none of the above

5. Solve.
   $7x - 14 = 91$
   a. 11
   b. 77
   c. 15
   d. 105

6. Estimate.
   $87.85
   16.14
   14.06
   + 9.28
   a. $116
   b. $126
   c. $127
   d. none of the above

7. Kai bought a $50 radio on sale for 35% off. What facts are given?
   a. regular price and discount rate
   b. regular price and sale price
   c. sale price and discount rate
   d. sale price and regular price

8. Evaluate.
   $100 - n^2$, if $n = 7$
   a. 51
   b. 49
   c. 93
   d. 149

9.
   | Brownsville | 26 miles |
   | Mt. Andrews | 78 miles |

   About how long would it take to ride a bike from Brownsville to Mt. Andrews at 25 miles per hour?
   a. 25 minutes
   b. 52 minutes
   c. 1 hour
   d. 2 hours

10. What is the length of the hypotenuse of the right triangle to the nearest whole number?

    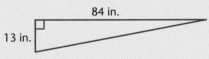

    a. 83 in.
    b. 85 in.
    c. 86 in.
    d. none of the above

84 Cumulative Test

CHAPTER 4

Ratios and Proportions

Christopher Councill
Calvert Day School

4.1 Ratios and Rates

Objective: to write ratios and rates

In the United States, there are about 80 telephones for every 100 people.

A **ratio** is a comparison of two numbers. The ratio that compares 80 to 100 can be written as follows.

80 out of 100   80 to 100   80:100 $\frac{80}{100}$

The ratio 80 to 100 can be written as a fraction in simplest form.

$\frac{80 \div 20}{100 \div 20} = \frac{4}{5}$ ▶ The GCF of 80 and 100 is 20.

More Examples

A. Change the ratio 30 out of 42 to a fraction in simplest form.

$\frac{30 \div 6}{42 \div 6} = \frac{5}{7}$ 5 out of 7

B. Change the ratio 63:9 to a fraction in simplest form.

$\frac{63 \div 9}{9 \div 9} = \frac{7}{1}$ 7:1

C. The ratio $\frac{135 \text{ revolutions}}{3 \text{ minutes}}$ compares the number of revolutions a record makes to the number of minutes it takes to make the revolutions. A ratio of two measurements having different units of measure is called a **rate**.

To find the number of revolutions the record makes each minute, simplify the rate so that the denominator is 1. A rate with a denominator of 1 is called a **unit rate**.

 = $\frac{45 \text{ revolutions}}{1 \text{ minute}}$ The unit rate is 45 revolutions per minute (45 rpm).

Try THESE

Write each ratio as a fraction in simplest form.

1. 4 out of 9
2. 1 to 1
3. 10:1
4. 4:2
5. 9 out of 12
6. 3 to 6

Write as a rate.

7. 100 miles in 4 hours
8. 24 pounds lost in 3 weeks

Write as a unit rate.

9. $\frac{\$430.00}{5 \text{ weeks}}$
10. $\frac{505 \text{ words}}{10 \text{ minutes}}$
11. $\frac{18 \text{ pounds}}{15 \text{ days}}$

Exercises

Write each ratio as a fraction in simplest form.

1. 6 to 16
2. 2 out of 16
3. 11 out of 11
4. 72:18
5. 63:105
6. 462 to 770
7. 0.2:5
8. $\frac{1}{4}$ to $\frac{1}{2}$
9. $\frac{1}{3}$ to $\frac{1}{9}$
10. $\frac{3}{4}$ to 10
11. 0.03:0.005
12. 24:2.4

Write as a rate.

13. $320.00 saved in 8 months
14. 225 kilometers in 5 hours
15. 12 revolutions in 0.5 seconds
16. 87.6 miles on 2.5 gallons

Write as a unit rate.

17. 88 calculators in 8 days
18. $3.84 for 3 gallons
19. 6 pounds gained in 8 weeks
20. 250 grams in 2.5 liters
21. 9.6 meters in 0.4 hours
22. $6.00 in 0.4 hours

Problem SOLVING

23. There are 15 girls and 12 boys in a class. What is the ratio of the number of boys to the total number of students in the class?
24. The Senate passed a bill by a ratio of 8 to 1. If 90 senators voted in all, how many voted for and how many voted against the bill?
25. A club has 15 members. The ratio of men to women in the club is 2:1. How many women are in the club?
26. Connie Klingler travels 243 miles in 4.5 hours. What is her hourly rate?
27. Jose paid $2.55 for a 15-minute long-distance telephone call. Fran paid $1.75 for a 10-minute call. Which person paid less per minute?
28. Caroline saves $45 in 4 months. At that rate, how long will it take her to save $180?

Test PREP

29. A 25-lb bag of Chow Chow puppy food costs $22.50. A 50-lb bag costs $42.50. How much money per pound would you save by buying the bag with the lower unit price?

    a. $0.20
    b. $0.50
    c. $0.05
    d. $0.10

4.2 Rate of Change

Objective: to find rates of change

During the day, the temperature rose 20°F. As the day progressed, the temperature rose until it reached its high for the day.

This change is an example of a **rate of change**. A rate of change describes how one quantity changes in relation to another.

In the graph to the right, you can see how the temperature changed from 8:00 A.M. to 4:00 P.M. The rate of change can be seen below.

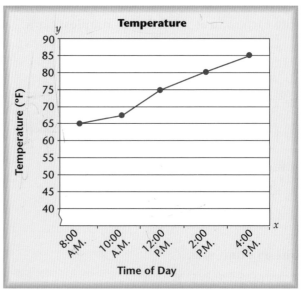

change of degrees during the day → $\dfrac{(85 - 65) \text{ degrees}}{8 \text{ hours}} = \dfrac{20 \text{ degrees}}{8 \text{ hours}} = 2.5$ **degrees per hour**

change in number of hours ↗

The graph above shows data points that are connected by line segments. A rate of change measures how fast a segment goes up or down when you read the graph from left to right.

You can find the rate of change by using the x- and y-coordinates.

> To find the rate of change, divide the change in the y-coordinates by the change in the x-coordinates.
>
> The rate of change between (x_1, y_1) and (x_2, y_2) is $\dfrac{y_2 - y_1}{x_2 - x_1}$.

A rate of change can be positive, negative, or zero.

Rate of Change	Positive (line slants up)	Negative (line slants down)	Zero (horizontal line)
Meaning	Increase	Decrease	No change
Graph	↗	↘	→

You can also find the rate of change by using a table.

Example

Find the rate of change from the table. Interpret this change.

Cakes Sold (x)	1	2	3	4
Dollars Earned (y)	$6	$12	$18	$24

$$\text{Rate of change} = \frac{\text{change in dollars earned}}{\text{change in number of cakes sold}} = \frac{24 - 6}{4 - 1} = \frac{18}{3} = \frac{6}{1}$$

The rate of change is $\frac{6}{1}$. You earned $6 for each cake sold.

Try THESE

State if the situation has a rate of change that is *positive, negative,* or *zero*.

1. the temperature of a cup of tea as it cools
2. the speed of a car as it accelerates
3. the speed of a car as it travels at a constant speed

Exercises

Use the table to the right to solve problems 1–5.

1. Find the rate of height change, in inches, from age 5 to age 7.
2. Find the rate of height change, in inches, from age 7 to age 10.
3. Between which two ages is the rate of height change greatest?
4. Make a graph of this data.
5. Is the rate of height change *positive, negative,* or *zero*?

Age (yr)	Height (in.)
5	45
6	47
7	51
8	53
9	54
10	56
11	59

Use the graph to the right to solve problems 6–9.

6. Find the rate of change in distance from 0 to 10 minutes.
7. Find the rate of change in distance from 20 to 30 minutes.
8. During which time period did the rate of change remain constant? How can you explain this?
9. During which time interval is the rate of change in distance the greatest?

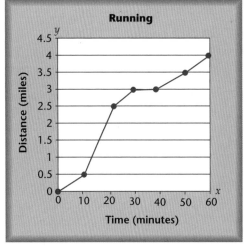

4.2 Rate of Change

4.3 Slope

Objective: to find the slope of a line from a graph and a table

A road sign may warn drivers of steep downgrades or upgrades. The sign at the right means that for every 5 feet, there is a rise of 1 foot.

The ratio of vertical distance to horizontal distance is called the **slope**. This ratio is sometimes referred to as the rise over the run. The change in the **rise** over the **run** is a constant rate of change.

In a coordinate system, the slope of a line is the ratio of the change in y to the corresponding change in x.

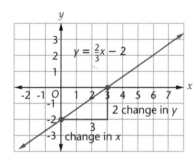

$$\text{slope} = \frac{\text{rise}}{\text{run}}$$

$$\text{slope} = \frac{\text{change in } y}{\text{change in } x} \quad \begin{array}{l}\text{vertical}\\ \text{horizontal}\end{array}$$

$$\text{slope} = \frac{2}{3}$$

$$\frac{0}{\text{change in } x} = 0 \quad \begin{array}{l}\text{The slope of a}\\ \text{horizontal line is 0.}\end{array}$$

$$\frac{\text{change in } y}{0} \;\blacktriangleright\; \text{no value} \quad \begin{array}{l}\text{A vertical line}\\ \text{has no slope.}\end{array}$$

> The change in y can be found by subtracting the y-coordinates.
>
> The change in x can be found by subtracting the corresponding x-coordinates.

You can find the slope of a line by looking at a graph.

Examples

A. Find the slope of the line.

Choose two points on the line. (-1, 1) and (2, 2)

Find the vertical change. $2 - 1 = 1$

Find the horizontal change. $2 - -1 = 3$

The slope of the line is $\frac{1}{3}$.

Since slope is a rate of change it can be positive (slant up), negative (slant down), or zero (horizontal).

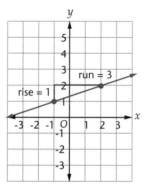

You can also find the slope of a line from a table.

B. Use the data in the table to find the slope. Then graph the data and the line.

x	-1	0	1	2
y	4	2	0	-2

Slope = $\dfrac{\text{change in } y}{\text{change in } x} = \dfrac{-2 + -4}{2 + -1} = \dfrac{-6}{3} = \dfrac{-2}{1} = -2$

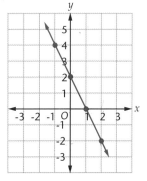

The slope of the line is -2.

Try THESE

Identify the rise and run of each graph.

1.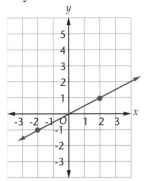

2.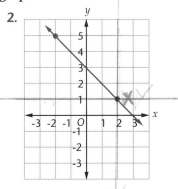

3.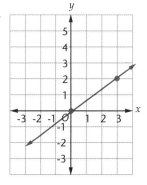

Exercises

Find the slope of each line.

1.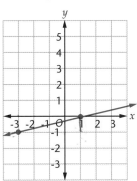

2.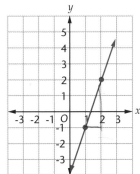

3.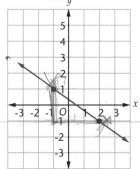

4.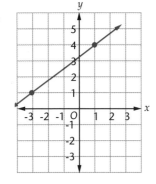
5.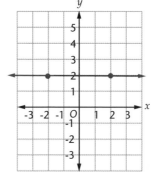
6.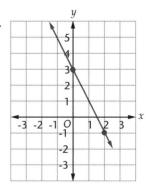

Find the slope of each line using the points given in each table.

7.
x	0	1	2	3
y	-4	-3	-2	-1

8.
x	0	2	4	6
y	6	5	4	3

9.
x	-4	-2	1	4
y	-2	0	2	4

10.
x	-1	0	1	2
y	-4	-1	2	5

Find the slope in problems 11–14. Use the graph at the right.

11. line k
12. line l
13. line m
14. line n

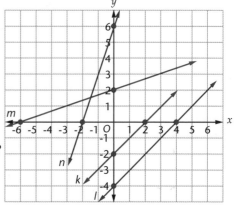

15. What is the relationship between the steepness of the graph of a line and the slope of the line?

16. What is the physical relationship between lines l and k?

17. What do you notice about the slopes of lines l and k?

Problem SOLVING

18. What is true about the slopes of parallel lines?

19. A ladder rests against a wall 20 feet up the wall. The foot of the ladder is 12 feet from the base of the wall. What is the slope of the ladder?

Constructed RESPONSE

20. A student found the slope of a line that crossed through the points (8, 7) and (4, 6) to be 4. This is not correct. What is the correct slope? How did the student find the incorrect slope? Explain.

Cumulative Review

Estimate.

1. $351 + 490$
2. $6{,}082 - 693$
3. $\$495 \times 57$
4. $55\overline{)436}$

5. $4.6 \bullet 2.7$
6. $8.2 - 6$
7. $48.19 \div 3.7$
8. $7.7 + 3.28$
9. $7\frac{1}{5} - 5\frac{3}{10}$
10. $\frac{3}{4} + \frac{5}{6}$
11. $\frac{7}{8} \bullet 5$
12. $\frac{4}{5} \div 3$

Find the GCF and LCM of each group of numbers.

13. 16, 40
14. 12, 36, 64
15. 2, 6, 10

Write as a fraction or mixed number in simplest form.

16. $\frac{21}{24}$
17. $\frac{6}{42}$
18. $\frac{37}{5}$
19. $\frac{38}{72}$
20. $\frac{94}{8}$
21. $\frac{39}{13}$

Solve.

22. $x + 7 = -9$
23. $12 = m - -8$
24. $y + 3\frac{1}{5} = 8\frac{2}{3}$
25. $7 = \frac{1}{2}c$
26. $r + 3.2 = 6.5$
27. $-7x = 4.2$
28. $\frac{2}{3}x = \frac{5}{6}$
29. $m - 1.18 = 1.58$

30. A pane of glass costs 69¢. What does it cost to replace the glass in 6 windows that have 4 panes each?

31. Bananas are on sale for $0.29 a pound. About how many pounds can Mr. Rodriquez buy for $2.00?

32. Kathy ran 5 miles each day for 5 days and 3 miles each day for 3 days. How many miles did she average each day?

33. Fala has $6\frac{1}{4}$ bags of plant food. She uses $1\frac{1}{4}$ bags each month. How many months will her supply of plant food last?

4.4 Conversions Using Dimensional Analysis

Objective: to use dimensional analysis

In science class, the teacher asked, "How many feet are in 1.5 miles?" To answer this question you need to know the relationship between feet and miles. You also need to know how to convert between units of measurement.

Common Units of Measure

Type	Length (customary)	Length (metric)	Capacity (customary)	Capacity (metric)	Weight (customary)	Mass (metric)
Unit	Inch (in.) Foot (ft) Yard (yd) Mile (mi)	Centimeter (cm) Meter (m) Kilometer (km)	Fluid ounce (fl oz) Cup (c) Pint (pt) Quart (qt) Gallon (gal)	Milliliter (mL) Liter (L)	Ounce (oz) Pound (lb) Ton (T)	Gram (g) Kilogram (kg)
Equivalent	1 ft = 12 in. 1 yd = 3 ft 1 mi = 5,280 ft	1 m = 100 cm 1 km = 1,000 m	1 c = 8 fl oz 1 pt = 2 c 1 qt = 2 pt 1 gal = 4 qt	1 L = 1,000 mL	1 lb = 16 oz 1 T = 2,000 lb	1 kg = 1,000 g

You choose the units you want to use to measure an object by the size of the object. What units should you use to measure your height or the distance from home to school?

Height—In the customary system, you should use feet. In the metric system, you should use meters.

Distance from home to school—In the customary system, you should use miles. In the metric system, you should use kilometers.

You can convert between units by using conversion factors, such as $\frac{2c}{1pt}$ or $\frac{1pt}{2c}$.

Conversion factors are rates equal to 1. You can analyze which rates to use by using **dimensional analysis**.

Examples

A. Use dimensional analysis to convert 3.5 cups to pints. Choose the conversion factor that will allow you to cancel the common units.

$3.5 \text{ cups} = \frac{3.5 \cancel{c}}{1} \cdot \frac{1 \text{ pt}}{2 \cancel{c}}$ Multiply by a conversion factor and cancel the common units.

$= \frac{3.5 \cdot 1}{2}$ Simplify.

$= 1.75 \text{ pints}$

In this example, the conversion factor $\frac{2c}{1pt}$ would not cancel the common unit of cups.

Sometimes you need to use more than one conversion factor.

B. A swimming pool is being filled up at a rate of 20 gallons per hour. Convert this to cups per minute.

$$\frac{20 \text{ gal}}{1 \text{ h}} = \frac{20 \text{ gal}}{1 \text{ h}} \cdot \frac{16 \text{ c}}{1 \text{ gal}} \cdot \frac{1 \text{ h}}{60 \text{ min}} \quad \text{Cancel out the common units.}$$

$$= \frac{20 \cdot 16 \cdot 1 \text{ c}}{1 \cdot 1 \cdot 60 \text{ min}} \quad \text{Simplify.}$$

$$= 5.3$$

The swimming pool is being filled up at a rate of 5.3 c/min.

Try THESE

Choose an appropriate customary unit for each measure.

1. the length of a piece of paper
2. the weight of a person

Choose an appropriate metric unit for each measure.

3. the capacity of a large glass
4. the distance from Baltimore to New York

Exercises

Choose an appropriate customary and metric unit for each measure.

1. the width of a penny
2. the capacity of a coffee mug
3. the mass of your Math book
4. the mass of a car
5. the width of a kitchen
6. the capacity of a swimming pool

Use dimensional analysis to convert each measure.

7. 5 ft = ☐ yd
8. ☐ pt = 12 c
9. 14.8 km = ☐ m
10. 550 lb = ☐ T
11. 2.8 yd = ☐ in.
12. ☐ qt = 15 gal

Use dimensional analysis to find an equal rate.

13. 60 mi/h = ☐ mi/min
14. 24 cm/s = ☐ m/s
15. 5 mi/h = ☐ ft/min

Problem SOLVING

16. Some tennis players can serve a ball at a speed of 85 mi/h. How many feet per minute is this?
17. Working out on an elliptical machine can help you burn 220 calories in 25 minutes. How many calories per minute are you burning? How many calories per hour are you burning?

4.5 Proportions

Objective: to solve proportions

A survey shows that 30 out of 50 people watch Channel 6 News. Suppose the survey represents a viewing area with 50,000 people. The ratio should be the same for the total population. Of the 50,000 people, 30,000 should watch Channel 6 News.

The two ratios, $\frac{30}{50}$ and $\frac{30,000}{50,000}$, name the same number, $\frac{3}{5}$.

The ratios are equivalent. An equation that states that two ratios are equivalent is called a **proportion**.

Two ratios form a proportion only if their **cross products** are equal.

$$\left. \frac{30}{50} = \frac{30,000}{50,000} \right\} \text{proportion}$$

$$\left. 30 \cdot 50,000 = 50 \cdot 30,000 \right\} \text{cross products}$$

$$1,500,000 = 1,500,000$$

Both cross products equal 1,500,000.

You can solve proportions by using cross products or mental math.

Examples

A. $\frac{d}{24} = \frac{15}{60}$

$d \cdot 60 = 24 \cdot 15$ Write the cross products.

$60d = 360$ Multiply.

$d = 6$ Solve for d.

B. Solve mentally.

$\frac{d}{24} = \frac{15}{60}$ $\frac{15}{60} = \frac{1}{4}$

$\frac{d}{24} = \frac{1 \cdot 6}{4 \cdot 6}$

$d = 6$

Try THESE

Solve.

1. What are the cross products for the proportion $\frac{3}{8} = \frac{9}{24}$?

2. Use cross products to determine if the ratios $\frac{36}{15}$ and $\frac{9}{5}$ form a proportion.

Solve each proportion.

3. $\frac{1}{5} = \frac{x}{35}$

4. $\frac{1}{3} = \frac{6}{s}$

5. $\frac{3}{a} = \frac{18}{24}$

6. $\frac{m}{3} = \frac{14}{21}$

Exercises

Write the cross products for each proportion.

1. $\dfrac{50}{25} = \dfrac{250}{125}$
2. $\dfrac{7}{16} = \dfrac{2.1}{4.8}$

Use cross products to see if each pair of ratios forms a proportion. Replace each ● with = (is equal to) or ≠ (is not equal to) to make a true statement.

3. $\dfrac{24}{4}$ ● $\dfrac{12}{2}$
4. $\dfrac{4}{9}$ ● $\dfrac{8}{18}$
5. $\dfrac{7}{5}$ ● $\dfrac{15}{10}$
6. $\dfrac{9}{1.35}$ ● $\dfrac{7}{1}$
7. $\dfrac{4}{6}$ ● $\dfrac{0.8}{1.2}$
8. $\dfrac{6}{8}$ ● $\dfrac{15}{20}$

Solve each proportion.

9. $\dfrac{1}{4} = \dfrac{n}{24}$
10. $\dfrac{1}{2} = \dfrac{a}{14}$
11. $\dfrac{r}{7} = \dfrac{8}{56}$
12. $\dfrac{5}{4} = \dfrac{25}{8}$
13. $\dfrac{8}{20} = \dfrac{25}{100}$
14. $\dfrac{h}{9.6} = \dfrac{5}{16}$
15. $\dfrac{2}{3} = \dfrac{1.2}{k}$
16. $\dfrac{r}{3} = \dfrac{8}{15}$
17. $\dfrac{2.4}{3.6} = \dfrac{d}{1.8}$
18. $\dfrac{s}{9.6} = \dfrac{7}{16}$
19. $\dfrac{x}{63} = \dfrac{9}{14}$
20. $\dfrac{x}{\frac{1}{2}} = \dfrac{4}{3}$
21. $\dfrac{9}{\frac{3}{4}} = \dfrac{x}{6}$
22. $\dfrac{4}{5} = \dfrac{8}{x+2}$
23. $\dfrac{3}{x+1} = \dfrac{12}{28}$

Problem SOLVING

24. A recipe uses 3 cups of flour to make 48 cookies. How much flour is needed to make 72 cookies?

25. If 3 bakers can bake 15 cakes in 1 hour, how long will it take 6 bakers to bake 20 cakes?

26. Lauren's car averages 33 miles for each gallon of gasoline. How many gallons are needed for a trip spanning 313.5 miles?

27. A 5-acre field has a yield of 280 bushels of wheat. What yield can be expected for a 42-acre field?

MiXeD REVIEW

Multiply or divide mentally.

28. $17 \div 10$
29. $8.435 \cdot 100$
30. $0.07 \cdot 10$
31. $264.5 \div 100$
32. $42.9 \div 1{,}000$
33. $1.0348 \cdot 1{,}000$
34. $649.7 \div 10^2$
35. $3.24 \cdot 10^3$
36. $41.267 \cdot 10^2$
37. $14.89 \div 10^1$
38. $56.09 \div 10^3$
39. $0.085 \div 10^0$

4.6 Problem-Solving Strategy: Using a Proportion

Objective: to solve problems using proportions

Ralph Mendoz is building a base for a television antenna. He needs 9 cubic feet of concrete. It takes 200 pounds of sand to make 4 cubic feet of concrete. How much sand does he need?

The ratios of sand to concrete must be the same. You know it takes 200 pounds of sand to make 4 cubic feet of concrete. You need to find how much sand it takes to make 9 cubic feet.

Set up a proportion. Let s represent the amount of sand needed.

$$\text{sand} \longrightarrow \frac{200}{4} = \frac{s}{9} \longleftarrow \text{sand}$$
$$\text{concrete} \longrightarrow \phantom{\frac{200}{4}} \phantom{\frac{s}{9}} \longleftarrow \text{concrete}$$

$$200 \cdot 9 = 4 \cdot s$$
$$1{,}800 = 4s$$
$$\frac{1{,}800}{4} = \frac{4s}{4}$$
$$450 = s$$

Mr. Mendoz needs 450 pounds of sand.

For 4 cubic feet of concrete, it takes 200 pounds of sand. For 9 cubic feet of concrete, it should take a little over 2 times as much sand or over 400 pounds of sand. The result is reasonable.

Set up a proportion for each of the following. Then solve each proportion.

1. It takes 260 pounds of gravel to make 4 cubic feet of concrete. It takes g pounds of gravel to make 9 cubic feet of concrete.

2. It takes 94 pounds of cement to make 4 cubic feet of concrete. It takes c pounds of cement to make 9 cubic feet of concrete.

3. 2 pounds yield 5 servings.
   n pounds yield 7 servings.

4. 9 gallons cost $12.15.
   n gallons cost $20.25.

98   4.6 Problem-Solving Strategy: Using a Proportion

Solve

The ratio of weight on Earth to weight on the Moon is 6:1. Find the equivalent weight for each of the following.

1. 120 pounds on Earth
2. 56 kilograms on the Moon
3. 155 kilograms on Earth
4. 2,000 pounds on the Moon
5. Find your equivalent weight on the Moon.

Solve. Use any strategy.

6. Wai Lui works 5 hours and earns $23.25. If he works 8 hours, how much does he earn?
7. Mr. Fisher plows a 10-acre field in $2\frac{1}{2}$ hours. How many acres can he plow in 14 hours?
8. A stew recipe uses 2 pounds of meat to make 12 servings. How much is needed to make 9 servings?
9. A piece of material $2\frac{1}{2}$ yards long costs $6.10. How much material can Joan buy for $4.27?
10. A recipe uses 3 cups of sugar to make 5 dozen cookies. How many cookies can be made with $1\frac{1}{4}$ cups of sugar?
11. A 165-mile trip takes 6 gallons of gas. How many gallons does a 240-mile trip take?
12. A bus driver estimates that it will take him 10 hours to drive 480 miles. After 4 hours, he has driven 200 miles. Is he on schedule?
★13. A commercial claims that 3 out of 5 dentists recommend Smile Toothpaste. At this rate, how many out of 350 dentists would *not* recommend Smile Toothpaste?
★14. From 9:00 A.M. to 3:00 P.M., Joe Ryan drives 280 kilometers. At the same rate, how far can he travel in 18 hours?

Mid-Chapter REVIEW

Write each ratio as a fraction in simplest form.

1. 20 to 12
2. 9 out of 15

Use dimensional analysis to find an equal rate.

3. 55 g = ☐ kg
4. 14 pt = ☐ c

Solve each proportion.

5. $\frac{21}{3} = \frac{63}{r}$
6. $\frac{a}{1.2} = \frac{3}{1.8}$

4.7 Similar Figures

Objective: to find missing measures in similar polygons

Rachel is on the yearbook staff at Wyandot Middle School. She must reduce the size of photographs to fit on the pages. The two photographs shown at the right have the same shape, but one is smaller than the other. Figures that have the same shape, but may differ in size are called **similar figures**.

7 in.
5 in.

h
2 in.

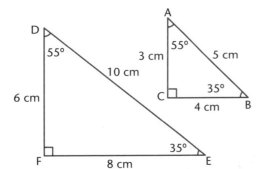

Triangles ABC and DEF are similar.

△ABC ~ △DEF

~ means "is similar to."

Compare the measures of the angles in △ABC with the measures of the corresponding angles in △DEF. What do you discover?

Next compare the measures of the corresponding sides. What do you discover?

> If two figures are similar, their corresponding angles are congruent and the measures of their corresponding sides are in proportion.

More Examples

A. Find the height of the reduced photograph above.

width of original → $\dfrac{5}{2} = \dfrac{7}{h}$ ← height of original
width of reduction →   ← height of reduction

$5 \cdot h = 2 \cdot 7$ Cross products are equal.

$5h = 14$

$\dfrac{5h}{5} = \dfrac{14}{5}$ Divide each side by 5.

$h = 2.8$ The height is 2.8 in.

You can determine whether two figures are similar by looking at the corresponding sides and angles.

B. Determine whether rectangle MNOP is similar to rectangle RSTU.

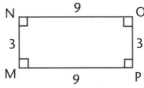
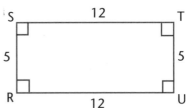

100   4.7 Similar Figures

Are the corresponding angles congruent?
Yes, they are all right angles.

Are the corresponding sides proportional?

$$\frac{MN}{RS} = \frac{3}{5} \qquad \frac{NO}{ST} = \frac{9}{12} = \frac{3}{4} \qquad \frac{OP}{TU} = \frac{3}{5} \qquad \frac{PM}{UR} = \frac{9}{12} = \frac{3}{4}$$

No, the corresponding sides are not proportional.

The figures are not similar because the corresponding sides are not proportional.

Try THESE

Complete. Refer to the similar figures at the right.

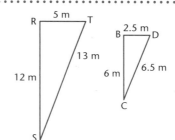

1. $\overline{TS}$ corresponds to _____.
2. The ratio of the measures of $\overline{BC}$ to $\overline{RS}$ is _____.

Exercises

Determine whether each pair of figures is similar. Explain your reasoning.

1.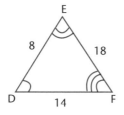

2.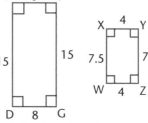

For each pair of similar figures, use a proportion to find the length of side x.

3.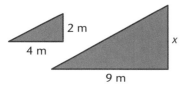

4.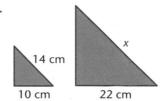

5.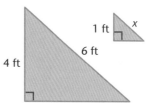

6.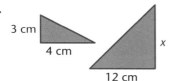

7.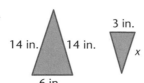

8.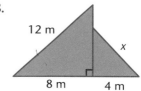

4.7 Similar Figures   **101**

Problem SOLVING

9. Find the length of the ladder on the right.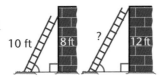

★10. Find the distance across the stream.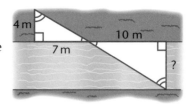

★11. A photo negative is 1.5 cm by 2.2 cm. The print is made to be 6 cm wide. How many times greater is the area of the print than the area of the negative?

Constructed RESPONSE

12. Describe characteristics common to two similar rectangles.

Mind BUILDER

Using a Stadia

A **stadia**, like the one at the right, can be used to find the height of an object.

Hold the stadia to your eye (strings at the far end) with an upright measuring stick in your line of sight. Back away from the object until the top and bottom of the object are in line with the strings. Now record the two readings where the strings appear to cross the stick. Notice that two similar triangles are formed.

1. Set up a proportion and find the height of the tree above.

2. Find an object, such as a building, tree, or flagpole. Use a stadia and a measuring stick to find the height of the object. Remember to measure the distances between you and the stick and between you and the object.

102   4.7 Similar Figures

Problem Solving

The Problem Problem

Lee is solving some problems. His progress per minute is in proportion to the number of problems left to be solved. The fewer problems he has left, the slower he works.

Lee solves the first problem in 5 minutes. The last problem takes 25 minutes to solve. How many problems did Lee have to solve?

Extension

How long did it take Lee to solve the problems?

4.8 Dilations

Objective: to graph dilations on a coordinate plane

The Tolbert family takes a vacation every year. This year, Mrs. Tolbert would like the children to help plan the trip. She uses a photocopier to enlarge her map, so everyone can read the markings.

Suppose a geometric figure is enlarged or reduced. The figure does not change its shape. However, it is altered in size. This is called a **dilation**. The dilation image is similar to the original figure.

The ratio found by comparing the distances from the center of a dilation to the original image's center is called the **scale factor**. In the figure at the right, $\triangle DEF$ is the dilation image of $\triangle ABC$. The distance from P to a point on $\triangle DEF$ is two times the distance from P to the corresponding point of $\triangle ABC$.

$PD = 2(PA)$

$PE = 2(PB)$

$PF = 2(PC)$

This dilation has center P and a scale factor of 2.

Another Example

Find the dilation image of $\triangle PQR$ with center C and a scale factor of $\frac{3}{4}$. Draw rays CP, CQ, and CR. Find X, Y, and Z so that $CX = \frac{3}{4}(CP)$, $CY = \frac{3}{4}(CQ)$, and $CZ = \frac{3}{4}(CR)$.

$\triangle XYZ$ is the dilation image of $\triangle PQR$.

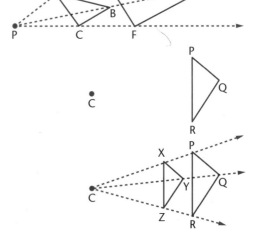

Copy the figure at the right. Then draw the dilation image of $\triangle JKL$ for the given scale factor and center.

1. scale factor = 2, center B

2. scale factor = $\frac{1}{2}$, center B

3. scale factor = 2, center A

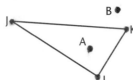

104   4.8 Dilations

Exercises

Graph each set of ordered pairs. Connect the points in order. On the same set of axes, draw the dilation image of each figure with the origin at the center and a scale factor of 2.

1. (0, 0), (0, 2), (4, 0)
2. (0, 0), (3, -3), (-2, -2)
3. (0, 0), (1, -2), (4, -3), (2, 3)

4–6. On the same sets of axes, draw the dilation images of the figures in Exercises 1–3 with a scale factor of $\frac{1}{2}$.

7. Compare the dilation images in Exercises 1–3 with the dilation images in Exercises 4–6. Are the figures similar?

Problem SOLVING

8. A 35-mm negative is 25 mm by 34 mm. A print made from the negative is 75 mm by 102 mm. What is the scale factor?

9. The viewfinder of a video camera is 1.2 cm high and 2.0 cm wide. What is the scale factor for a video screen that is 54 cm high and 90 cm wide?

10. Gina uses a magnifying glass to look at bugs. A caterpillar that is 2 inches long appears 4.5 inches long through the glass. What is the scale factor?

11. A state map has insets of the state capital and the downtown area of a large city. One inset is 4 inches wide, and the other is 3 inches wide. On the map, the regions are $\frac{1}{2}$ inch wide. What are the scale factors?

12. A map has a scale of 1 cm to 600,000 m. The distance between two peaks on the map is 8.4 cm. What is the actual distance in kilometers?

Mixed REVIEW

Rename each fraction or mixed number as a terminating or repeating decimal.

13. $\frac{4}{5}$
14. $\frac{15}{4}$
15. $6\frac{2}{3}$
16. $\frac{12}{5}$

Rename each decimal as a fraction or mixed number in simplest form.

17. 5.2
18. 0.38
19. 9.15
20. 3.06

4.9 Scale Drawings

Objective: to solve problems involving scale drawings

A **scale drawing** is used to represent something that is too large or too small to be conveniently drawn in actual size. A map is a scale drawing.

Mrs. Dorris lives in Terre Haute, Indiana. She wants to drive to Indianapolis. She can use the map to find the actual distance from Terre Haute to Indianapolis.

On the map, 1 centimeter represents 50 kilometers. The map distance between the two cities is about 2.0 centimeters.

Set up a proportion to find the actual distance d.

$$\text{map} \rightarrow \frac{1}{50} = \frac{2.0}{d} \leftarrow \text{map}$$
$$\text{actual} \rightarrow \phantom{\frac{1}{50}} \phantom{\frac{2.0}{d}} \leftarrow \text{actual}$$

$$1 \cdot d = 50 \cdot 2.0$$
$$d = 100$$

The actual distance from Terre Haute to Indianapolis is about 100 kilometers.

Try THESE

Find the actual distance between the cities. Use the map of Indiana shown above. Remember that on this map, 1 cm = 50 km. The measurement between each pair of cities has already been provided.

1. Gary, New Albany, 6.3 cm
2. Anderson, Muncie, 0.5 cm
3. Kokomo, Lafayette, 1.0 cm
4. Fort Wayne, Richmond, 3.0 cm
5. New Albany, Richmond, 3.1 cm
6. Kokomo, Anderson, 1.1 cm

Exercises

Find the actual length of each trip. Use the map on p. 106.

1. Terre Haute to Indianapolis to Bloomington
2. Fort Wayne to Richmond to New Albany
3. Lafayette to Kokomo to Anderson to Muncie

Scale drawings may be shown on grid paper. Use the scale drawing of the living room to answer each question.

4. What is the scale?
5. What are the actual dimensions of the room in feet?
6. What are the actual dimensions of the sofa in inches?
7. What are the actual dimensions of each chair in inches?
★ 8. How many square yards of carpet would it take to cover the floor?

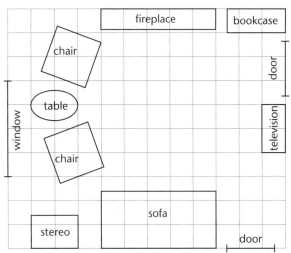

Scale: $\frac{1}{4}$ inch = 18 inches

Problem SOLVING

9. On a scale drawing of a building, 1 inch represents 5 feet. The actual width of a window is $3\frac{1}{2}$ feet. How wide should the window be on the scale drawing?

10. A water tower that is 144 feet high has a diameter of 16 feet. What is the ratio of the diameter to the height?

★ 11. The actual width of a car is 4 feet 8 inches. On a scale drawing, 1 inch represents 0.7 feet. How wide is the car on the scale drawing?

12. Find the dimensions of your bedroom. Choose an appropriate scale, and draw a scale model of your bedroom.

4.10 Problem-Solving Application: Using Indirect Measurement

Objective: to measure indirectly using similar triangles

Royalton Construction Company is clearing land for a new housing development. Any tree under 6 meters tall is cut down. Felicia Loomis uses a method involving similar triangles to find the height of a tree. This is called **indirect measurement**.

Miss Loomis holds a meterstick perpendicular to the ground. Then she measures its shadow and the shadow of the tree. This is known as shadow reckoning. The tree, the meterstick, and their shadows form two sides of similar triangles from which a proportion can be written.

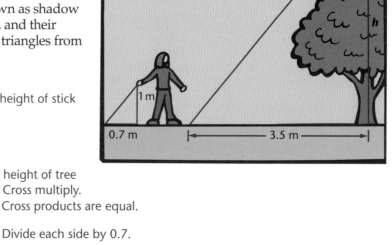

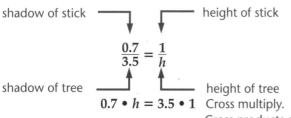

$0.7 \cdot h = 3.5 \cdot 1$ Cross multiply.
Cross products are equal.

$\dfrac{0.7h}{0.7} = \dfrac{3.5}{0.7}$ Divide each side by 0.7.

$h = 5$ The tree is 5 meters tall.

Another Example

Corinne is 5 feet tall and casts a 2-foot shadow. Find the height of a building that casts a 26-foot shadow.

$\dfrac{2}{26} = \dfrac{5}{h}$

$2 \cdot h = 26 \cdot 5$

$\dfrac{2h}{2} = \dfrac{130}{2}$

$h = 65$ The height of the building is 65 feet.

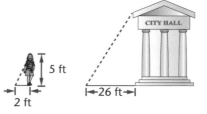

Try THESE

The shadow of a meterstick measures 0.8 m. The lengths of the shadows of trees measured at the same time are given below. Use a proportion to find the height of each tree to the nearest tenth of a meter.

1. 3 m
2. 9 m
3. 5.1 m

Solve

1. Find the length of Kingly Lake.

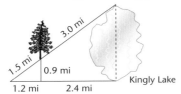

2. Find the height of the brace.

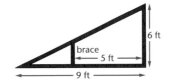

3. A flagpole casts a 9-meter shadow. At the same time, a 2-meter signpost casts a 2.6-meter shadow. How tall is the flagpole?

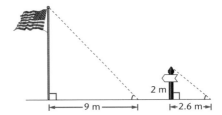

4. A tower casts a 32-meter shadow. At the same time, a 2-meter pole nearby casts a shadow of 8 meters. How high is the tower?

5. When Ayani stands 3 feet from a lamppost, his shadow is 4 feet long. If Ayani is 6 feet tall, how tall is the lamppost?

6. Angelita is 1.5 meters tall and casts a 0.75-meter shadow. At the same time, a nearby flagpole casts a 3-meter shadow. How tall is the flagpole?

7. Pam is 57 in. tall, and Rory is 63 in. tall. They are standing by a lamppost. Pam's shadow is 76 in. long. How long is Rory's shadow?

★8. Find the length of Cedar Lane.

9. Find the distance across the river.

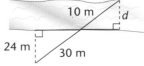

Constructed RESPONSE

10. Explain how you would use indirect measurement to determine the height of a building.

4.10 Problem-Solving Application: Using Indirect Measurement   **109**

4.11 Tangent, Sine, and Cosine Ratios

Objective: to find the tangent, sine, and cosine of angles

The measures of the sides of right triangles form special ratios, as defined below.

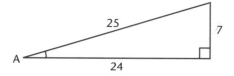

$$\text{sine of } \angle A = \frac{\text{measure of the side opposite } \angle A}{\text{measure of the hypotenuse}} \quad \text{or} \quad \sin A = \frac{7}{25} = 0.280$$

$$\text{cosine of } \angle A = \frac{\text{measure of the side adjacent to } \angle A}{\text{measure of the hypotenuse}} \quad \text{or} \quad \cos A = \frac{24}{25} = 0.960$$

$$\text{tangent of } \angle A = \frac{\text{measure of the side opposite } \angle A}{\text{measure of the side adjacent to } \angle A} \quad \text{or} \quad \tan A = \frac{7}{24} \approx 0.292$$

$\approx$ means "is approximately equal to."

You can use the table on p. 326 to find the sine, cosine, and tangent ratios for an angle.

Examples

A. Use the table to find tan 53°.

1. Find 53° in the angle column.

Angle	sin	cos	tan
52°	0.7880	0.6157	1.2799
53°	0.7986	0.6018	1.3270

2. Find the corresponding reading in the tan column.

So, tan 53° is approximately 1.3270.

This means that the side opposite the 53° angle is about 1.3 times greater than the side adjacent to it.

B. Suppose you know that the length of the side opposite an angle is twice the side adjacent to it. In other words, the tangent of the angle is 2 or 2.0000. You can use the table to find the measure of the angle.

2. Find the corresponding readings in the angle column.

Angle	sin	cos	tan
62°	0.8829	0.4695	1.8807
63°	0.8910	0.4540	1.9626
64°	0.8988	0.4384	2.0503
65°	0.9063	0.4226	2.1445

Start here.

1. Find the values that are close to 2.0000 in the tan column.

The angle is between 63° and 64°. Since 2.0000 is closer to 1.9626 than to 2.0503, the angle is about 63°.

Try THESE

Express each ratio to three decimal places.

1. sin J
2. cos J
3. tan J
4. sin L
5. cos L
6. tan L

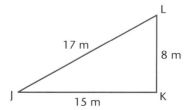

Exercises

Use the table on p. 326 to find the value of each ratio.

1. sin 30°
2. cos 30°
3. tan 38°
4. sin 51°
5. cos 45°
6. cos 72°
7. tan 34°
8. cos 0°

Use the table on p. 326 to find the measure of each angle.

9. sin M = 0.8660
10. tan N = 0.5000
11. cos F = 0.7071
12. tan E = 0.5774
★13. sin A = $\frac{1}{2}$
★14. cos T = $\frac{4}{5}$

Problem SOLVING

15. Can the value of the sine ratio ever be greater than 1? Why or why not? Remember that the sine is the ratio of the measure of the opposite side to the measure of the hypotenuse.

★16. What is the greatest possible value of the tangent ratio? Remember that the tangent is the ratio of the measure of the opposite side to the measure of the adjacent side.

17. What is the measure of angle A?

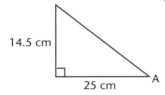

18. Find the measures of angle A and angle C.

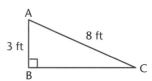

MiXeD REVIEW

Solve each equation.

19. $z + 7 = 35$
20. $s - 14 = 52$
21. $17 = a + 9$
22. $6y = 48$
23. $\frac{k}{4} = 52$
24. $20.8 = 16t$
25. $7.3 = b - 2.6$
26. $1.7 = \frac{n}{13}$
27. $27 = 2c + 3$
28. $\frac{p}{7} - 4 = 35$
29. $23 = \frac{r}{12} + 15$
30. $6x - 14 = 40$

4.11 Tangent, Sine, and Cosine Ratios

4.12 Problem-Solving Application: Using Tangent, Sine, and Cosine

Objective: to solve problems using tangent, sine, and cosine

Jean Newton works for the WROY radio station in Royalton. She needs to verify the height of the broadcasting tower. Suppose the tower casts a shadow 40 meters long. The angle measures 47°. The tangent ratio may be used to find the height of the tower.

$\tan 47° = \dfrac{x}{40}$ ← opposite side / adjacent side

$1.0724 \approx \dfrac{x}{40}$ Use the table on p. 326.

$1.0724 \cdot 40 \approx \dfrac{x}{40} \cdot 40$

$42.896 \approx x$

The tower is about 43 meters tall.

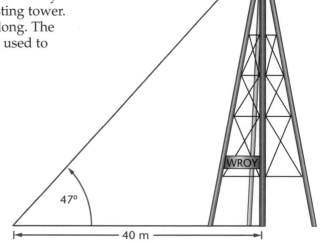

Another Example

You can find the degree measures of angles A and B using either the sine ratio or the cosine ratio.

$\sin A = \dfrac{5}{13}$ $\cos A = \dfrac{12}{13}$

≈ 0.3947 ≈ 0.9231

(triangle: 13 in. hypotenuse, 5 in., 12 in.)

Using the table on p. 326, you see that by either method, the degree measure of angle A is approximately 23°. Therefore, the measure of angle B is approximately 90° − 23°, or 67°.

Try THESE

Find the missing angles and sides of each triangle.

1.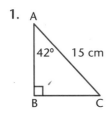

2.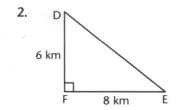

3.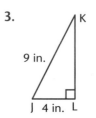

Solve

Solve. Round your answers to the nearest unit.

1. How high is the kite?

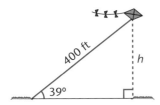

2. How long is the guide wire?

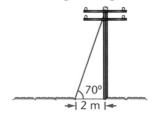

3. Lola and Ed are flying kites in Crockett Park. When Lola is 30 meters from Ed, Lola's kite is directly over Ed. The kite string makes an angle of 58° with the ground. How high is the kite?

4. The path of a cable car connecting Piney Peak and Smokey Peak rises 20 meters for every 100 meters of horizontal distance. What is the measurement of the angle the cable makes with the horizontal distance?

5. A tent has a center pole that is 6 feet high and a floor 8 feet wide. Find the measure of the angle the tent side forms with the ground.

6. **Research** Find out what a sextant is and how it can be used to measure angles.

★7. For safety, the base of a ladder should be placed at a distance of one-fourth of the ladder's length from the wall. What is the measure of the angle the ladder forms with the ground?

Use the table on the right to solve each problem.

8. Write the ratio of Milwaukee's games won to games lost as a fraction in simplest form.

9. Write the ratio of Baltimore's games won to the total number of games played. Compare this to the number in the Pct. column for Baltimore.

★10. Detroit has 54 more games left in the season. Suppose their win-loss record for the rest of the season is in proportion to their present record. How many more games will they win?

AMERICAN LEAGUE STANDINGS

East	W	L	Pct.	Games Behind	Last 10 Games	Streak
Cleveland	70	42	0.625	—	6–4	Lost 1
New York	63	47	0.573	6	8–2	Won 7
Detroit	60	48	0.556	8	4–6	Lost 2
Boston	56	54	0.509	13	4–6	Lost 4
Baltimore	55	54	0.505	$13\frac{1}{2}$	4–6	Lost 1
Milwaukee	50	58	0.463	18	6–4	Won 1
Toronto	37	74	0.333	$32\frac{1}{2}$	5–5	Won 2

4.12 Problem-Solving Application: Using Tangent, Sine, and Cosine

Chapter 4 Review

Language and Concepts

Choose the correct term to complete each sentence.

1. A _____ compares two numbers and can be written as follows: 2 out of 3, 2 to 3, 2:3, or $\frac{2}{3}$.

2. A _____ is an equation in the form $\frac{a}{b} = \frac{c}{d}$, which states that two ratios are equivalent.

3. A ratio that has two measurements with different units of measure is called a _____.

4. If two figures are similar, their corresponding angles are _____.

5. When something is too large or too small to be drawn in actual size, a _____ is used.

6. The ratio of vertical distance to horizontal distance is called the _____.

7. A _____ is an image that has been altered in size but not in shape.

slope
congruent
rise
cross product
proportion
rate
dilation
ratio
scale drawing
similar

Skills and Problem Solving

Write each ratio as a fraction in simplest form. Section 4.1

8. 5 out of 6
9. 3 to 6
10. 15:10

Write as a unit rate. Section 4.1

11. 450 miles in 9 hours
12. $12.00 in 1.5 hours

Use the table to the right to solve each problem. Section 4.2

13. Find the rate of height change from month 1 to month 3.

14. Is the rate of height change *positive, negative,* or *zero*?

Time (month)	Height (inches)
1	5
2	9
3	11

Find the slope of each line. Section 4.3

15.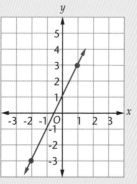

16.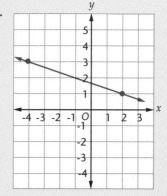

Use dimensional analysis to find an equal rate. Section 4.4

17. 4.8 yd = ☐ in.
18. $7\frac{1}{2}$ c = ☐ qt
19. 25 oz = ☐ lb

Solve each proportion. Section 4.5

20. $\frac{1}{3} = \frac{n}{12}$
21. $\frac{18}{48} = \frac{6}{y}$
22. $\frac{2}{3} = \frac{x}{0.6}$
23. $\frac{8}{2.5} = \frac{10.4}{c}$

Use the figure at the right to solve each problem. Section 4.7

24. DE = 4 in., CE = 6 in., BD = 2.5 in.,
    AC = _____

25. AC = 8 cm, BD = 6 cm, BE = 10.5 cm,
    AE = _____

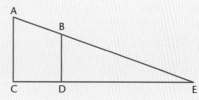

Graph (1, 3), (2, 4), and (4, 1). Section 4.8

26. Connect the points in order. On the same set of axes, draw the dilation image with the origin at the center and a scale factor of 3.

On a scale drawing of a boat, 1 cm represents 3.5 m.
Scale drawing measurements are given below. Find the actual measurement of each of the following. Section 4.9

27. length, 6.7 cm
28. width, 4.8 cm

Express each ratio to three decimal places. Section 4.11

29. sin J
30. tan K

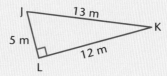

Use the table on p. 326 to find the measure of each angle. Section 4.11

31. tan A = 0.3443
32. cos B = 0.1908

Solve. Sections 4.6, 4.10

33. A recipe calls for $2\frac{1}{2}$ cups of flour to make 3 dozen cookies. How much flour is needed to make 90 cookies?

34. Luis saves $29.00 in 4 months. At that rate, how long will it take him to save $116.00?

35. A tree casts a shadow of 24 feet, while Ralph casts a shadow of 8 feet. If Ralph is 6 feet tall, how tall is the tree?

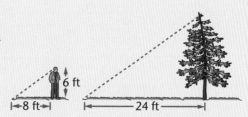

Chapter 4 Test

Write each ratio as a fraction in simplest form.

1. 14 to 28
2. 21:9
3. 8 out of 18

Find the slope of each line using the points in each table.

4.
x	-1	0	1	2
y	2	3	4	5

5.
x	0	1	2	3
y	1	-1	-3	-5

Use dimensional analysis to find an equal rate.

6. 15 in./min = ☐ ft/h
7. 25.75 cm/s = ☐ m/min

Solve each proportion.

8. $\frac{1}{a} = \frac{11}{22}$
9. $\frac{3}{8} = \frac{y}{12}$
10. $\frac{4}{3} = \frac{10}{w}$
11. $\frac{6}{5} = \frac{f}{17.5}$

Set up a proportion. Then solve each proportion.

12. 2 gallons cost $2.38.
    n gallons cost $10.71.

13. 6 packages cost 57¢.
    4 packages cost n cents.

On a scale drawing of a calculator, 1 cm represents 4.5 cm. Scale drawing measurements are given below. Find the actual measurement of each.

14. width, 1.8 cm
15. length, 3.1 cm

Solve.

16. Name the corresponding angles and sides of the similar triangles.

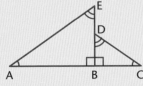

17. Express sin T, cos T, and tan T to three decimal places.

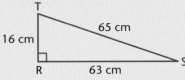

Use the table on p. 326 to find each value.

18. sin 71°
19. cos 15°
20. tan 34°

Solve.

21. Mr. Thomas won an election by a ratio of 6 to 2. His opponent received 2,300 votes. How many votes did Mr. Thomas receive?

22. Maria drives 484 kilometers in $5\frac{1}{2}$ hours. At the same rate, how far can she travel in 12 hours?

23. What is the distance across the pond?

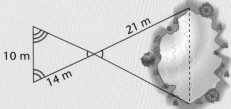

24. How high is the building?

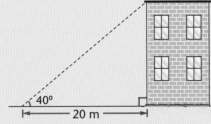

Change of Pace

Angle of Elevation

To see the top of a peak or an object in the sky, a person must look up rather than look straight ahead. An angle of elevation is formed by the line of sight and a horizontal line.

You can make a device called a **hypsometer** to measure the angle of elevation. Then you can use the tangent ratio to find the height of the object.

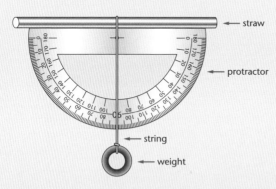

Tie a weight to a 10-inch string. Then tie the string around the center of a drinking straw. Tape the straw to the base of a large protractor. Be sure the string is placed at the center mark on the protractor.

Use the hypsometer to find the height of a tree.

- Look through the straw to the top of the tree.
- Read the angle measure where the string crosses the protractor.
- Measure the horizontal distance from you to the tree.
- Use the tangent ratio to find the height.
- Add the distance from the ground to your eyes to find the actual height of the tree.

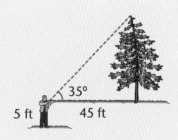

$$\tan 35° = \frac{x}{45}$$

$$0.7002 = \frac{x}{45}$$

$$0.7002 \cdot 45 = x$$

$$31.5 = x$$

The tree is 31.5 + 5, or 36.5 ft high.

Solve.

1. A chimney casts a shadow 75 feet long when the angle of elevation of the Sun is 41°. How tall is the chimney?

2. A road rises 38 feet vertically over a horizontal distance of 540 feet. What is the angle of elevation of the road?

Cumulative Test

1. What is the ratio 18:22 changed to a fraction in simplest form?
   a. $\frac{18}{22}$
   b. $\frac{9}{11}$
   c. $\frac{11}{9}$
   d. none of the above

2. Solve.
   $\frac{a}{5} + 6 = 6.5$
   a. 0.1
   b. 0.5
   c. 2.5
   d. 62.5

3. Evaluate. $32 + 7 \cdot 2$
   a. 26
   b. 32
   c. 19
   d. 46

4. How long would it take a horse, running 30 miles per hour, to run 2 miles? Use $d = rt$.
   a. 4 minutes
   b. 6 minutes
   c. 40 minutes
   d. 60 minutes

5. What property is shown by $2.1 + 5.03 = 5.03 + 2.1$?
   a. associative
   b. commutative
   c. distributive
   d. identity

6. How is the sentence "Six more than x is nine" written as an equation?
   a. $6 = x + 9$
   b. $6 + 9 = x$
   c. $6 + x = 9$
   d. none of the above

7. Larry Weaver's heating bills have increased each month. In December, the bill was $85.63. It increased by $9.85 in January and another $5.52 in February. What was the total amount he paid for December, January, and February?
   a. $101.00 b. $272.26
   c. $282.11 d. none of the above

8. What is the total number of passengers?
   a. 2,860 million
   b. 3,140 million
   c. 5,952 million
   d. 6,052 million

Passengers by Vehicle Type

Motor Bus 4,406 million

Rail 1,546 million

9. Which of the following is equivalent to $(2^4)(2^5)$?
   a. 2^9
   b. 2^{20}
   c. 4^9
   d. 4^{20}

10. Which facts are needed to find the amount Mr. Craig paid?
    1. Mr. Craig bought three shirts.
    2. Each shirt cost $21.95.
    3. The sales tax rate is 6.5%.
    4. Mr. Craig gave the clerk $80.00.
    a. 1 and 3
    b. 2 and 3
    c. 1, 2, and 3
    d. 1, 2, 3, and 4

CHAPTER 5

Percents

Ashley Gilmore
Maryland

5.1 Fractions, Decimals, and Percents

Objective: to convert between fractions, decimals, and percents

Shawn works at Angelo's Pizza House. To determine the amount of pepperoni and sausage to order, he marks a square blue for sausage and red for pepperoni when each is ordered. He finds that 53 out of 100 choices are for pepperoni. What percent is this?

A **percent** is a ratio that compares some number to 100. Since percent means "per hundred," the ratio $\frac{53}{100}$ can be expressed as 53%.

You can write a fraction as a percent in one of the following ways.

$\frac{4}{5}$

$\frac{4 \times 20}{5 \times 20} = \frac{80}{100}$ Find an equivalent fraction with a denominator of 100.

$\frac{4}{5} = 80\%$

$\frac{2}{3}$

$\frac{2}{3} = \frac{x}{100}$ Set up a proportion.

$3x = 200$ Cross products are equal.

$x = 66\frac{2}{3}$ Divide each side by 3.

$\frac{2}{3} = 66\frac{2}{3}\%$

$\frac{5}{6}$

$\frac{5}{6}$ Divide.

$6\overline{)5.00}$ $0.83\frac{1}{3}$

$\frac{5}{6} = 83\frac{1}{3}\%$

Examples

You can also write a percent as a fraction in simplest form.

A. $14\% = \frac{14}{100}$

$= \frac{\cancel{14}^{7}}{\cancel{100}_{50}}$ or $\frac{7}{50}$

So, $14\% = \frac{7}{50}$.

B. $0.75\% = \frac{0.75}{100}$

$= \frac{0.75}{100} \times \frac{100}{100}$

$= \frac{\cancel{75}^{3}}{\cancel{10,000}_{400}}$ or $\frac{3}{400}$ So, $0.75\% = \frac{3}{400}$.

Since a percent is a ratio that has a denominator of 100, all percents can be expressed as decimals. Likewise, all decimals can be expressed as percents.

C. Rename 15% as a decimal.

$15\% = \frac{15}{100}$

Say: fifteen hundredths

Write: 0.15

D. Rename 125% as a decimal.

$125\% = \frac{125}{100} = \frac{100}{100} + \frac{25}{100} = 1.25$

Say: one hundred twenty-five hundredths

Write: 125% = 1.25

E. Rename 0.28 as a percent.

Say: twenty-eight hundredths

Write: $\frac{28}{100}$

Since percent means "per hundred," $\frac{28}{100} = 28\%$.

F. Rename 0.259 as a percent.

$\frac{259 \div 10}{1,000 \div 10} = \frac{25.9}{100} = 25.9\%$

Say: 259 thousandths

Write: 0.259 = 25.9%

Try THESE

Rename each fraction or mixed number as a percent.

1. $\frac{1}{5}$
2. $\frac{3}{4}$
3. $\frac{9}{20}$
4. $2\frac{7}{20}$

Exercises

Rename each fraction or mixed number as a percent.

1. $\frac{4}{7}$
2. $1\frac{3}{8}$
3. $1\frac{2}{9}$
4. $1\frac{9}{16}$
5. $2\frac{3}{7}$

Write each percent as a fraction or mixed number in simplest form.

6. 45%
7. 30%
8. 8%
9. 5%
10. 27%
11. 160%
12. 139%
13. $87\frac{1}{2}$%
14. $16\frac{2}{3}$%
15. $11\frac{1}{9}$%
16. $144\frac{4}{9}$%
17. $143\frac{3}{4}$%

Write each decimal as a percent.

18. 0.24
19. 0.006
20. 0.102
21. 1.24
22. 1.07
23. 0.9
24. 0.205
25. 0.709

Write each percent as a decimal.

26. 70%
27. 36%
28. 41.5%
29. 70.6%
30. 137%
31. 245%
32. 0.9%
33. 0.6%
34. 0.2%
35. $133\frac{1}{3}$%
36. $1\frac{1}{6}$%
37. $6\frac{5}{6}$%

Replace each ● with <, >, or = to make a true statement.

38. 1.5 ● 100%
39. 0.495 ● 1%
40. 6.1 ● 610%
41. 0.35 ● 3.6%

Problem SOLVING

42. About two out of every seven students attend the Allen Junior High track meet. What percent of the student body attends the track meet?

43. If $\frac{1}{3}$ is equivalent to $33\frac{1}{3}$%, what percent is equivalent to $\frac{2}{3}$?

44. On a city block, $\frac{4}{5}$ of the people subscribe to the newspaper. What percent subscribe to the newspaper?

45. The product of two numbers is $3\frac{1}{4}$. One factor is $3\frac{5}{7}$. Write the other factor as a decimal and as a percent.

5.1 Fractions, Decimals, and Percents

5.2 Finding Percents Mentally

Objective: to compute percents mentally

During a clearance sale at Eastside Audio, 29 out of the 41 stereos were sold. About what percent were sold?

You can estimate the percent by renaming a nearly equivalent fraction.

THINK $\frac{29}{41}$ is about $\frac{30}{40}$ or $\frac{3}{4}$.

$\frac{3}{4} = \frac{75}{100}$ or 75%   So about 75% of the stereos were sold.

Fraction Equivalents	
$33\frac{1}{3}\% = \frac{1}{3}$	$66\frac{2}{3}\% = \frac{2}{3}$
$25\% = \frac{1}{4}$	$75\% = \frac{3}{4}$
$20\% = \frac{1}{5}$	$60\% = \frac{3}{5}$
$16\frac{2}{3}\% = \frac{1}{6}$	$83\frac{2}{3}\% = \frac{5}{6}$
$12\frac{1}{2}\% = \frac{1}{8}$	$87\frac{1}{2}\% = \frac{7}{8}$

> Knowing common fraction-percent equivalents will help you estimate with percents.

More Examples

A. Estimate the percent that is shaded.

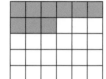

9 out of 30

$\frac{9}{30}$ is about $\frac{10}{30}$ or $\frac{1}{3}$.

$\frac{1}{3} = 33\frac{1}{3}\%$

About $33\frac{1}{3}\%$ is shaded.

B. Estimate 12% less than $300.00.

12% is about 10% or $\frac{1}{10}$.

$\frac{1}{10}$ of $300.00 is $30.00.

300 − 30 = 270

$300.00 less 12% is about $270.00.

C. Estimate 0.5% of $988.00.

THINK

0.5% is half of 1%.

988 is about 1,000.

1% means "1 out of 100."

1,000 ÷ 100 = 10

$\frac{1}{2}$ of 10 is 5.

0.5% of $988.00 is about $5.00.

D. Estimate 123% of 8.

THINK

123% is more than 100%.

So, 123% of 8 is greater than 8.

123% is about 125%.

125% = 1.25 or $1\frac{1}{4}$

$8\left(1 + \frac{1}{4}\right) = 8 + 2$

$= 10$

123% of 8 is about 10.

Try THESE

Estimate the percent.

1. 6 out of 23
2. 3 out of 40
3. 41 out of 58
4. $\frac{31}{25}$

Exercises

Match each percent to an equivalent fraction.

1. $66\frac{2}{3}\%$
2. 75%
3. 50%
4. 10%
5. $33\frac{1}{3}\%$
6. 40%
7. 20%
8. 1%
9. 100%

a. $\frac{3}{4}$ b. $\frac{2}{5}$ c. $\frac{2}{3}$
d. $\frac{1}{2}$ e. $\frac{1}{5}$ f. $\frac{1}{10}$
g. $\frac{1}{1}$ h. $\frac{1}{100}$ i. $\frac{1}{3}$

State the fraction, mixed number, or whole number you would use to estimate.

10. 27%
11. 65%
12. 97%
13. 36%
14. 0.3%
15. $4\frac{1}{2}\%$
16. 157%
17. $10\frac{1}{5}\%$
18. $\frac{3}{8}\%$
19. 32.4%
20. 24.98%
21. $45\frac{4}{5}\%$
★22. 267.2%
★23. 0.15%
★24. $118\frac{1}{2}\%$

Estimate.

25. 78% of 20
26. 24% of 84
27. 9% of 32
28. 65% of 85
29. 48% of $23.95
30. 98% of $5.50
★31. 1.5% of 135
★32. 125% of 79
★33. 0.6% of 205
★34. 24 is what percent of 50?
★35. 30 is 11% of what number?

Problem SOLVING

36. Alice plays basketball. In one game, she shoots 16 free throws and makes 81.25% of them. About how many free throws does she make?

37. In a survey of 550 people, 62% like Tasty Pizza. About how many people like Tasty Pizza?

38. In a school with 582 students, 36% are in the eighth grade. About how many eighth graders are there in the school?

Constructed RESPONSE

39. Mrs. Johnson estimates that a 15% tip on a $58.50 restaurant bill is $5. Is this a reasonable estimate? Explain.

Mixed REVIEW

Compute.

40. $1\frac{1}{2} + \frac{3}{4}$
41. $2\frac{1}{3} - 1\frac{5}{7}$
42. $14\frac{7}{8} + 5\frac{1}{5}$
43. $6 - 3\frac{4}{5}$
44. $\frac{1}{4} \times 26$
45. $34\frac{2}{3} - 31\frac{1}{2}$
46. $10 \div \frac{3}{4}$
47. $45\frac{1}{2} \div \frac{1}{3}$

5.2 Finding Percents Mentally

5.3 Percents and Proportions

Objective: to use proportions to solve percent problems

A tire store sold 250 tires in one week. Out of the 250 tires sold, 84% were radial tires. How many radial tires were sold? You can use a proportion to solve this problem.

$\frac{Part}{Whole} = \frac{Percent}{100}$ or $\frac{a}{b} = \frac{p}{100}$, where a is the part, b is the whole, and p is the percent.

Examples

A. Find 84% of 250.

$\frac{a}{b} = \frac{p}{100} \longrightarrow \frac{a}{250} = \frac{84}{100}$ Replace b with 250 and p with 84.

$a \bullet 100 = 250 \bullet 84$ Find the cross products.

$\frac{100a}{100} = \frac{21{,}000}{100}$ Multiply.

$a = 210$ Divide.

210 radial tires were sold.

You can also use the percent proportion to find the percent or the base.

Type	Example	Proportion
Find the Percent	10 is what percent of 50?	$\frac{10}{50} = \frac{p}{100}$
Find the Whole	10 is 20% of what number?	$\frac{10}{b} = \frac{20}{100}$
Find the Part	What number is 20% of 50?	$\frac{a}{50} = \frac{20}{100}$

B. What percent of 80 is 52?

$\frac{52}{80} = \frac{p}{100}$

$52 \bullet 100 = 80 \bullet p$

$\frac{5{,}200}{80} = \frac{80p}{80}$

$65 = p$

52 is 65% of 80.

C. 35% of what number is 21?

$\frac{21}{b} = \frac{35}{100}$

$21 \bullet 100 = 35 \bullet b$

$\frac{2{,}100}{35} = \frac{35b}{35}$

$60 = b$

35% of 60 is 21.

Try THESE

Identify the part, whole, and percent in each of the following.

1. 25% of 80 is 20.
2. 35 is 50% of 70.
3. 10 out of 15 is $66\frac{2}{3}$%.

Exercises

Write a percent proportion to solve each problem. Then solve. Round your answers to the nearest tenth, if necessary.

1. Find 40% of 60.
2. What number is 50% of 126?
3. What percent of 90 is 9?
4. 72 is 6% of what number?
5. 0.3% of what number is 9?
6. 16% of what number is 1,920?
7. What number is 15% of 20?
8. What percent of 50 is 12.5?
9. $6\frac{1}{2}$% of 900 is what number?
10. What number is 500% of 10?
11. 120% of what number is 540?
12. 50 is 6.25% of what number?

Problem SOLVING

13. Mrs. Miller sells a house for $179,000. If she earns a commission of 6%, how much money does she earn?
14. Mr. Jameson buys a suit for 80% off the regular price. He pays $180 for the suit. What is its regular price?
15. It rained 40% of the days in April. How many days did it rain in April?
16. Dan receives $196.84 out of his total earnings of $259.00. What percent of his earnings does he receive?

Mind BUILDER

Mental Math

Inflation occurs when a rise in total spending is more than the rise in production and service.

Inflation Rate	Buying Power of $100 in 5 Years	Buying Power of $100 in 10 Years
4%	$81.54	$66.48
6%	$73.39	$53.86
9%	$62.40	$38.94

Solve using mental math.

1. At 9% inflation, what is the buying power of $1,000.00 in 10 years?
2. At 4% inflation, what is the buying power of $1.00 in 5 years?
3. **Make Up a Problem**
   Write another problem you can solve mentally using the table.

5.4 Percent Equations

Objective: to solve problems using the percent equation

John's school requires 40 hours of community service. John spent 36 hours volunteering at a hospital. What percent of the requirement has he fulfilled? You can use an equation to solve this problem.

The percent proportion $\frac{Part}{Whole} = Percent$ can be transformed into an equation by multiplying both sides by the whole.

The percent equation is *Part = Percent • Whole*.

Remember that when you write a percent as a decimal, you divide the percent by 100, and when you write a decimal as a percent, you multiply by 100.

Examples

A. 36 is what percent of 40?

Part = Percent • Whole	percent equation
36 = *P* • 40	Replace *Part* with 36 and *Whole* with 40.
36 = 40*P*	Multiply.
0.9 = *P*	Divide.

36 is 90% of 40.

You can also find the part and the whole using the percent equation.

Type	Example	Proportion
Find the Percent	10 is what percent of 50?	10 = *P* • 50
Find the Whole	10 is 20% of what number?	10 = 0.20 • *W*
Find the Part	What number is 20% of 50?	*P* = 0.20 • 50

B. 14% of what number is 49?

49 = 0.14 • *W*

49 = 0.14*W*

350 = *W*

14% of 350 is 49.

C. What number is 75% of 48?

P = 0.75 • 48

P = 36

36 is 75% of 48.

Try THESE

Identify the part, whole, and percent in each of the following.

1. $33\frac{1}{3}\%$ of what number is 96?
2. Find 15% of 750.
3. 584 is what percent of 1,250?
4. 89.25 is 105% of what number?

Exercises

Write a percent equation to solve each problem. Then solve. Round your answers to the nearest tenth, if necessary.

1. 820 is 20% of what number?
2. What is 4.2% of 68?
3. 17,000 is what percent of 8,500?
4. 25 is $62\frac{1}{2}\%$ of what number?
5. 68 is 2% of what number?
6. What is 0.05% of 48?
7. 25 is what percent of 350?
8. 225% of what number is 900?
9. What is 127% of 250?
10. What is $66\frac{2}{3}\%$ of 324?
★11. $\frac{4}{5}$ is what percent of $\frac{5}{4}$?
★12. 0.5% of what number is 16?

Problem SOLVING

13. A solution is $2\frac{1}{2}\%$ distilled water. How many gallons of distilled water are there in 250 gallons of solution?

14. On a 40-question test, Peggy answers 34 correctly. What percent of the questions does she answer correctly?

Constructed RESPONSE

15. Cassie's salary is $125 per week plus an 8% commission on all sales. How much must she sell to earn $200 per week? Explain, and show all work.

Mid-Chapter REVIEW

Solve.

1. What number is 20% of 45?
2. 95% of 135 is what number?
3. What percent of 8 is 6?
4. 21 is what percent of 56?
5. 57.6 is 48% of what number?
6. 15% of what number is 90?
7. Rename 20% as a decimal and as a fraction in simplest form.
8. Out of 100 students, 12 are on the student council. What percent are not on the student council?

Problem Solving

Super Fly

Two bike riders that are 175 miles apart begin traveling toward each other at noon. One travels at 20 miles per hour; the other travels at 15 miles per hour. Also at noon, a fly begins flying between the riders, starting at the front of the slower bike. The fly travels at 20 miles per hour and can change direction without losing any time.

How far will the fly travel before the bicycles meet?

Extension

If the fly travels at 25 mph, how far will it travel before the riders meet?

Cumulative Review

Copy and complete each pattern.

1. 13, 27, 41, ____, 69, ____, ____
2. 400, 200, ____, ____, 25, ____
3. $10\frac{1}{2}, 9\frac{3}{4}, 9, 8\frac{1}{4}$, ____, ____, ____
4. 30, 29, 27, 24, 20, ____, ____, ____

Evaluate each expression if $a = 3$, $b = 4$, and $c = 2.5$.

5. $6b - c$
6. $a + b^2$
7. $\frac{2b}{c}$
8. $a^3 - \sqrt{b}$

Solve each equation.

9. $m + 3 = 9$
10. $8 + r = 15\frac{1}{2}$
11. $x - 5 = 7$
12. $y - 2\frac{1}{2} = 8$
13. $13m = 156$
14. $39 = 3y$
15. $\frac{x}{6} = 9$
16. $63 = \frac{y}{7}$
17. $6m = 4\frac{1}{2}$
18. $5y - 8 = 32$
19. $\frac{c}{2} - 8 = 12$
20. $2x + 2\frac{1}{2} = 4\frac{1}{2}$

Compute. Write each answer in simplest form.

21. $\frac{9}{16} + \frac{11}{16}$
22. $\frac{11}{15} - \frac{2}{15}$
23. $\frac{7}{8} \cdot \frac{24}{49}$
24. $\frac{4}{5} \div \frac{3}{8}$
25. $\frac{7}{9} - \frac{3}{5}$
26. $\frac{5}{8} + \frac{5}{12}$
27. $\frac{7}{12} \div 7$
28. $1\frac{3}{8} \cdot 10$
29. $8\frac{5}{6} + 3\frac{5}{12}$
30. $4\frac{5}{8} - 1\frac{11}{12}$
31. $1\frac{7}{8} \cdot 2\frac{5}{6}$
32. $4\frac{1}{3} \div 1\frac{5}{6}$

Solve each proportion.

33. $\frac{4}{n} = \frac{24}{60}$
34. $\frac{c}{9} = \frac{3}{2}$
35. $\frac{1}{2} = \frac{x}{15}$
36. $\frac{2}{3} = \frac{m}{100}$

Solve.

37. It costs $272 per month to feed a bear. The City Zoo has five bears. How much does it cost to feed the bears for 1 year?

38. An empty milk crate weighs 1 pound. Each book weighs 4.25 pounds. What is the total weight when the crate is filled with ten books?

39. A recipe for party mix calls for 2 cups of corn cereal for 10 servings. How much corn cereal is needed for 25 servings?

40. On a scale drawing of a house, 1 inch represents 3 feet. If the dimensions of the kitchen are 9 feet by $13\frac{1}{2}$ feet. What are the dimensions on the scale drawing?

5.5 Percent of Change

Objective: to find the percent of increase and decrease

Andrew Evans had a paper route with 35 customers. A new apartment building was added to the paper route. He now has 49 customers. The percent of his customers increased.

You can express an increase or a decrease in a quantity using a percent. The percent that a quantity increases or decreases from its original amount is called the **percent of change**.

Percent of increase is when the new amount is greater than the original, and percent of decrease is when the new amount is less than the original.

$$\text{percent of change} = \frac{\text{amount of change}}{\text{original amount}}$$

Examples

A. Find the percent of change from 35 customers to 49 customers.

- Step 1: Subtract to find the amount of change. $49 - 35 = 14$ The route increased by 14 customers.

- Step 2: Set up a ratio to compare the amount of change to the original number of customers.

$$P = \frac{14}{35} \quad \frac{\text{The amount of change is 14.}}{\text{The original amount is 35.}}$$

$$= 0.40 \quad \text{Divide. Write as a percent.}$$

The percent of increase is 40%.

B. A stereo system is reduced from $380 to $323. Find the percent of decrease.

- Step 1: Subtract to find the amount of decrease. $380 - 323 = 57$ The price is decreased by $57.

- Step 2: Set up a ratio to compare the amount of change to the price of the stereo.

$$P = \frac{57}{380} \quad \frac{\text{The amount of change is 57.}}{\text{The original amount is 380.}}$$

$$= 0.15 \quad \text{Divide. Write as a percent.}$$

The percent of decrease is 15%.

Try THESE

Find the percent of increase or decrease.

1. a $40.00 radio increased to $44.00
2. a $450 oven decreased to $396
3. 40 books to 45 books
4. 170 pounds to 136 pounds

Exercises

Find the percent of increase or decrease.

1. a $60.00 lamp decreased to $43.20
2. a $1,300 sofa decreased to $650
3. 69¢ lettuce increased to 92¢
4. a 30-student class decreased to 28
5. 20 points per game to 24 points
6. a 6-minute mile decreased to 5 minutes and 24 seconds

Problem SOLVING

7. A calculator used to sell for $29.00. It now sells for $25.52. What is the percent of decrease?

8. At the start of the football season, Jorge weighed 150 pounds. At the end of the season, he weighed 138 pounds. What was the percent of change in his weight?

9. The first time she does yard work, Felicia mows the yard in $2\frac{1}{2}$ hours. The next time, she mows it in 2 hours and 5 minutes. What is the percent of change?

10. One year, a store sold 120 stereos. The next year, 150 stereos were sold. What is the percent of increase?

11. The value of an antique table increased from $320.00 to $360.00 in 3 years. What is the percent of change?

Constructed RESPONSE

12. A pair of hiking boots originally cost $42.00. The price was increased 25%. The boots were then sold with the price reduced $\frac{1}{4}$. What was the final sale price? Explain.

Test PREP

13. Of 84 sofas on display at a furniture store, about 12% have matching pillows. How many sofas have matching pillows?

    a. 10
    b. 11
    c. 12
    d. 13

5.5 Percent of Change

5.6 Markup, Discount, and Sales Tax

Objective: to solve problems involving markups, discounts, and sales tax

John Clark marks up the price of merchandise in his sporting goods store 60% more than what he pays for it. **Markup** is the percent of increase in price to cover expenses and make a profit.

$$\text{percent of markup} = \frac{\text{markup}}{\text{store's cost}} \quad \text{or} \quad \text{markup} = \text{percent of markup} \cdot \text{store's cost}$$

Examples

A. Mr. Clark's cost for a pair of roller skates is $20. What is the selling price of the roller skates if the markup is 60%?

The markup is equal to the percent times the store's cost.

$m = 0.60 \cdot 20$ Replace *percent* with 0.60 and *store's cost* with 20.

$m = 12$ Multiply.

$\$12 + \$20 = \$32$ Add the markup to the store's cost to find the selling price.

Mr. Clark is selling the roller skates for $32.

A **discount** is the amount of reduction from the regular price. The regular price minus the discount is the **sale price**.

B. Kelly wants a bike from Mr. Clark's sporting goods store. The bike usually sells for $198. If she buys it at 20% off, how much will she save? What is the sale price?

The discount is equal to the percent times the original amount.

$d = 0.20 \cdot 198$ Replace *percent* with 0.20 and *original amount* with $198.

$d = 39.60$ Multiply.

$\$198 - \$39.60 = \$158.40$ Subtract the discount from the original price to find the sale price.

The price of the bike on sale is $158.40.

C. If the sales tax rate is 5%, what is the total cost of the bike?

Let s = sales tax.

$s = 0.05 \cdot \$158.40$

$s = \$7.92$

To find the total cost of the bike, add the sale price and the sales tax.

total cost = $158.40 + $7.92

= $166.32

The total cost with the sales tax is $166.32.

Try THESE

Find the amount of the discount and the sale price for each item.

1. $40 jacket, 20% off
2. $300 earrings, 30% off
3. $8.49 CD, $\frac{1}{3}$ off
4. $799 sofa, 10% off

Exercises

Find the selling price for each item to the nearest cent.

1. weight set, $48.65
   50% markup
2. running shoes, $20.00
   80% markup
3. sleeping bag, $12.50
   100% markup
4. baseball bat, $6.30
   90% markup

Problem SOLVING

5. Mr. Clark's cost for a basketball is $15. The markup is 50%. What is the selling price?
6. Larry buys a $13 book and an $8 calendar. Each item is on sale for 20% off. What is the total cost of the items? The sales tax rate is 6%. What is the total cost with the sales tax?
7. Marie bought a sweater on sale for $20. The regular price was $25. What was the discount rate?
8. A ring is on sale for $105. It is 25% off its original price. What was its original price?
9. The selling price of a baseball glove at Pedro's Sporting Goods is $27.90. The percent of markup is 80%. How much did the store pay for the glove?
10. Josie bought a car for $10,955. The sales tax rate where she lives is 5.5%. What is the total cost of the car?

Constructed RESPONSE

11. Luisa has $25. Can she buy a $25 shirt at 40% off and an $18 pair of sandals at $\frac{1}{2}$ off? Show all work, and explain.

Test PREP

12. A book was $15.00 and is now $10.00. What is the discount rate?

    a. 25% off
    b. 15% off
    c. $33\frac{1}{3}$% off
    d. 36% off

5.7 Simple and Compound Interest

Objective: to find simple and compound interest

Sang Ko borrows $700 to buy a trumpet. He is charged 12.75% simple interest per year. How much interest is Sang Ko charged for 2 years?

Interest is money that is earned or paid for the use of money. Banks charge interest when they loan money. You are paid interest when you deposit money into their bank.

Simple Interest Formula

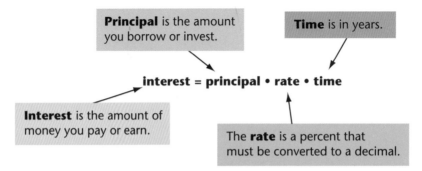

Examples

A. How much interest is Sang Ko charged for 2 years?

$I = prt$	Use the simple interest formula.
$I = 700 \cdot 0.1275 \cdot 2$	Replace p with 700, r with 0.1275, and t with 2.
$I = 178.5$	Multiply.

The simple interest for 2 years is $178.50.

B. What is the total amount Sang Ko paid back to the bank?

He borrowed $700 and was charged $178.50 in interest, so to find the total amount repaid, you should add.

$700 + $178.50 = $878.50

Sang Ko paid $878.50 back to the bank.

Most banks pay compound interest on savings accounts. Compound interest is added to an account after certain intervals. In this way, interest earns interest. Common intervals for compounding interest are semiannually, quarterly, and daily.

If you deposit $100.00 in a savings account that pays 6% interest compounded quarterly, you can find the principal plus interest as follows.

	principal •	rate •	time	=	interest
	p •	r •	t	=	I
First 3 months	100 •	0.015 •	1	=	$I \rightarrow 1.50$
	$p = 100 + 1.50 = \$101.50$				
Second 3 months	101.50 •	0.015 •	1	=	$I \rightarrow 1.5225$
	$p = 101.50 + 1.52 = \$103.02$				
Third 3 months	103.02 •	0.015 •	1	=	$I \rightarrow 1.5453$
	$p = 103.02 + 1.55 = \$104.57$				
Fourth 3 months	104.57 •	0.015 •	1	=	$I \rightarrow 1.568$
	$p = 104.57 + 1.57 = \$106.14$				

> *Quarterly* means "4 times in one year." Divide the annual interest rate by 4 to find the quarterly interest rate.
>
> $6\% \div 4 = 1.5\%$

The principal plus interest after 1 year is $106.14.

Try THESE

Find the simple interest to the nearest cent.

1. $10,000 at 13% for 6 months
2. $2,500 at 12% for $2\frac{1}{2}$ years

Exercises

Find the total amount in the account.

1. $650 at 3.5% for 5 years
2. $5,000 at 4.2% for 18 months
3. $115,000 at $4\frac{2}{5}$% for 3 years
4. $56 at 4% for 36 months
5. $400 at 5% compounded semiannually for 3 years

Problem SOLVING

6. Martha borrows $2,400. What is the total amount she pays back in 2 years at a simple interest rate of $14\frac{1}{2}$%?

7. Hannah wants to invest some money so that in 4 years she will be able to buy a new car. If she wants to earn $1,500 in interest at a rate of 5%, how much should she invest?

Constructed RESPONSE

8. Josephine invests $5,000 in an account. After 3 years, she has $5,975 in her account. What is the interest rate on the account? Show all work, and explain.

5.8 Problem-Solving Strategy: Guess and Check

Objective: to solve problems using guess and check

Emily decides to get a loan to buy a computer. She budgets $125 per month for the loan payment. The loan is for 1 year at 15% interest. How much can Emily afford to pay for the computer?

You must find the cost of a computer Emily can afford. You know the interest rate, the length of the loan, and Emily's maximum payment.

Multiply the budgeted monthly payment by 12 months. Then guess and check amounts close to this, including the 15% interest. By multiplying each guess by 115%, you get the full amount plus the interest.

125 • 12 = 1,500

Emily can spend a total of $1,500 for the computer and interest on the loan. Now guess and check to see what the total cost is for several prices of computers.

Guess	Check	
$1,400	$1,400 • 115% = $1,610	too high
$1,200	$1,200 • 115% = $1,380	too low
$1,300	$1,300 • 115% = $1,495	

Use each result to help you make the next guess.

Emily can afford a computer that costs about $1,300.

The solution is reasonable, since $1,300 at 15% interest is nearly $1,500.

Solve using guess and check.

1. Yana wants to buy a coat. He can afford $10.00 monthly payments for 1 year. The interest rate is 18%. How much can Yana afford to pay for the coat?

2. Cheryl has 6 coins in her pocket. The total value of the coins is $1.30. How many quarters, dimes, nickels, and pennies does she have?

Solve

Use any strategy.

1. Joey was ill. He lost 13 pounds. Now he weighs 117 pounds. What was the percent of decrease in his weight?

2. Mr. Cox is four times as old as his son. In 18 years, he will be twice as old as his son. What are their ages now?

3. Ms. West gets a 7% discount by paying her rent before the third day of each month. She pays $288.30 on May 1. How much does she save by paying early?

4. Ms. Amity bought a $130.00 suit and a $40.00 blouse, both for 25% off. The sales tax was 5.5% (based on the sale price). How much did Ms. Amity pay?

5. Find a digit for each letter so the problem is correct. Each letter represents a different digit.

    $$\begin{array}{r} ABCDE \\ \times 4 \\ \hline EDCBA \end{array}$$

6. Paper cups can be purchased in packages of 40 or 75. Atepa buys 7 packages and gets 350 cups. How many packages of 75 does she buy?

7. Ted bought a pair of $18.95 slacks and a $15.00 shirt, both for 30% off. What was his total cost (without tax)?

★8. Lucas has an average score of 83% on three tests. What must his score be on a fourth test to raise his average to 87%?

Mixed Review

Evaluate.

9. $8 \cdot 3 \div \sqrt{16} - 2$

10. $3.6 + 7.2 \div 4$

11. $(3^3 + 6) \cdot \sqrt{25}$

12. $(20 \cdot 6) - (11 \cdot 7)$

Translate each sentence into an equation.

13. The quotient of s and 4 is 5.

14. Three more than a number is six.

15. The sum of 4 times a number and 5 is 37.

Chapter 5 Review

Language and Concepts

Choose the correct term to complete each sentence.

1. A(n) _____ is a ratio that compares some number to 100.
2. Percent times whole is equal to _____.
3. To find the percent of increase or decrease, you compare the amount of change to the _____.
4. A(n) _____ is a decrease in price.
5. _____ is the term used for the percent of increase in price a store uses to cover expenses and make a profit.
6. Interest that is added to an account after certain intervals so the interest also earns interest is called _____.

percent
part
original
markup
discount
interest
compound interest

Skills and Problem Solving

Rename each fraction or mixed number as a percent. Section 5.1

7. $\frac{7}{8}$
8. $\frac{9}{5}$
9. $\frac{2}{5}$
10. $\frac{3}{8}$
11. $1\frac{3}{50}$

Rename each percent or decimal as a fraction or mixed number in simplest form. Section 5.1

12. 25%
13. $33\frac{1}{3}$%
14. 8%
15. 0.012
16. 1.05

Estimate the percent. Section 5.2

17. 12 out of 25
18. 9 out of 26
19. 16 out of 50
20. 28 out of 295

Solve. Sections 5.3–5.4

21. What number is 25% of 500?
22. 20% of 65 is what number?
23. What percent of 12 is 6?
24. 17 is what percent of 51?
25. 120 is 60% of what number?
26. 50% of what number is 46?
27. $83\frac{1}{3}$% of 48 is what number?
28. What is 0.6% of 59?
29. 250 is what percent of 100?
30. What percent of 400 is 2?
31. 1.2% of what number is 0.06?
32. 450% of what number is 49.5?

Find the percent of increase or decrease for each of the following. Section 5.5

33. a $50.00 vase decreased to $37.50
34. 15 minutes to 25 minutes

Find the sales tax for each of the following items. Then find the total cost of the item. Section 5.6

35. coat, $70.00
    tax rate, 5%
36. bicycle, $129.00
    tax rate, $4\frac{1}{2}$%
37. book, $4.95
    tax rate, 6%

Find the selling price for each item to the nearest cent. Section 5.6

38. picture, $45.00
    70% markup
39. shoes, $20.00
    45% markup
40. swing set, $58.00
    110% markup

Find the interest to the nearest cent. Section 5.7

41. $650 at 12% for 2 years
42. $1,250 at 10% for 3 months

Solve. Section 5.8

43. The Express Coffee Company had expenses of $273,509 and sales of $300,527 last year. What was the percent of profit based on sales?

44. Leroy has some dogs and birds. He counts all the heads and gets 10. He counts all the feet and gets 34. How many dogs and how many birds does he have?

Chapter 5 Test

Rename each decimal, fraction, or mixed number as a percent.

1. 0.37
2. 0.061
3. $\frac{31}{100}$
4. $2\frac{3}{10}$
5. $\frac{19}{25}$

Estimate.

6. 9% of 11
7. 35% of 150
8. 19% of 250
9. 250% of 48

Solve.

10. What number is 35% of 120?
11. 28% of 70 is what number?
12. 30 is what percent of 250?
13. What percent of 32 is 16?
14. 2.25 is 50% of what number?
15. 40% of what number is 35.8?
16. 5% of 50 is what number?
17. What number is $66\frac{2}{3}$% of 276?
18. What percent of 2.6 is 26?
19. 50.4 is what percent of 16,800?
20. 105% of what number is 37.8?
21. 1.08 is 9% of what number?
22. What number is 250% of 48?
23. What percent of 39 is 40?

Find the percent of increase or decrease for each of the following.

24. 40 tickets to 64 tickets
25. a $440 canoe decreased to $396

Find the sale price for each item.

26. $29.95 jeans, 25% off
27. $11.50 album, 15% off
28. $288.00 bike, $\frac{1}{3}$ off

Find the selling price for each of the following items.

29. blanket, $21.00 75% markup
30. album, $4.70 90% markup
31. mirror, $86.00 50% markup

Find the principal plus interest to the nearest cent.

32. $560 at 8% for 1 year
33. $4,000 at $9\frac{1}{4}$% compounded monthly for 6 months

Solve.

34. Tickets for the play are $2.50 for adults and $1.50 for students. Dario buys 8 tickets for $15.00. How many adult tickets does he buy?
35. Abby's mother lends her $600 for 4 years at 6% annual interest. How much does Abby owe her mother at the end of 4 years?

Change of Pace

Egyptian Mathematics

The oldest Egyptian writings were in a pictorial script called hieroglyphics. Developed before 3000 B.C., these special symbols were used to write counting numbers.

Number	1	10	100	1,000	10,000	100,000	1,000,000
Egyptian Symbol							
Name	Vertical Staff	Heel Bone	Rope Core	Lotus Flower	Pointing Finger	Burbot Fish	Surprised Man

The values of the symbols were added to find the value of the number.

"Twelve thousand three hundred fourteen" was written as follows.

Find the values of each number represented below.

Hieroglyphics	Numerical Values
|||||||	_____
∩∩||||| ∩∩||||	_____
???∩∩||| ∩∩	_____
⚱⚱??|	_____
⌒⚱⚱⚱???|||| ⚱⚱⚱???|||	_____

Can you find a pattern in the numbers? Look carefully! When you find the pattern, you will notice the error the scribe made when he copied this set of hieroglyphics. Which hieroglyphic number does not fit the pattern? How would you correct it?

Cumulative Test

1. Which numeral represents 365 thousandths?
   a. 0.0365
   b. 0.365
   c. 3.650
   d. 36.5

2. Solve.
   $\frac{4}{5} = \frac{x}{2}$
   a. 1
   b. 1.3
   c. 1.6
   d. 2.5

3. 60 is 80% of what number?
   a. 48
   b. 75
   c. 120
   d. 140

4. Estimate.
   7.04 • 38.6
   a. 210
   b. 280
   c. 2,800
   d. 28,000

5. Chad's quiz scores are 19, 15, 16, and 16. What must he score on the next quiz to have an average score of 17?
   a. 16
   b. 17
   c. 18
   d. 19

6. What is 8% of 15?
   a. 0.12
   b. 1.2
   c. 12
   d. 120

7. In the drawing below, which describes the relationship of Spring Street and Long Street?
   a. parallel
   b. perpendicular
   c. skew
   d. vertical

8. Two angles of a triangle measure 32° and 43°. What is the measure of the third angle?
   a. 15°
   b. 90°
   c. 105°
   d. 180°

9. Sam bought some stamps. He gave half the stamps to his brother and used $\frac{1}{3}$ of the remaining stamps. He now has 12 stamps. How many did he buy?
   a. 24
   b. 36
   c. 40
   d. none of the above

10. Wendy plays basketball. In one game she makes 9 field goals, or 60% of her shots. Which of the following conclusions is correct?
    a. Wendy took 100 shots and made 60 of them.
    b. Wendy missed 4 shots.
    c. Wendy took 15 shots.
    d. You cannot tell how many shots Wendy took.

CHAPTER 6

Polygons and Transformations

Donnaviv Foussier
Dubai

6.1 Angle Relationships

Objective: to explore and classify angles

Bridges are classified according to the way they are constructed. The bridge shown at the left is an example of an arch bridge. Likewise, angle pairs are classified according to the way they are formed.

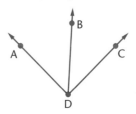

Two **angles** are **adjacent** when they have a common side, the same **vertex**, and do not overlap.

∠ADB and ∠BDC are adjacent angles.

Perpendicularity, right angles, and 90° measurements are all in the same classification. Lines, rays, or segments that intersect at right angles are perpendicular.

In the figure at the right, the mark inside the angle (⌐) indicates that ∠B is a **right angle**. It is also true that $\overline{AB}$ is perpendicular to $\overline{BC}$ and ∠B = 90°.

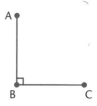

More Examples

Angle pairs can also be classified by the sum of their angles.

A. Two angles are **complementary** if the sum of their measures equals 90°.

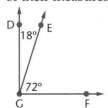

m∠DGE + m∠EGF = 18° + 72° = 90°

∠DGE and ∠EGF are complementary.

B. Two angles are **supplementary** if the sum of their measures equals 180°.

m∠I + m∠L = 115° + 65° = 180°

∠I and ∠L are supplementary.

C. Intersecting lines form pairs of supplemental angles and **vertical angles**.

Angles can be named by numbers.
∠1 and ∠3 are vertical angles.
∠2 and ∠4 are vertical angles.

Vertical angles have the same measure.
m∠1 = m∠3
m∠2 = m∠4

D. By joining the corners of any triangle, a very important angle relationship is revealed.

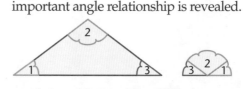

m∠1 + m∠2 + m∠3 = 180°

> The sum of the measures of the angles of a triangle equals 180°.

Exercises

Use the figure at the right to complete each of the following.

1. ∠QTS and ■ are right angles.
2. ∠QTR and ■ are complementary.
3. ∠QTN and ■ are supplementary.
4. If ∠NTR measures 148°, then ∠RTS measures ■.
5. ∠PTN is adjacent to ■.

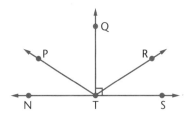

Use the figure at the right to complete each of the following.

6. ∠1 and ■ are vertical angles.
7. ∠3 and ■ are vertical angles.
8. If m∠1 = 40°, then m∠4 = ■.
9. If m∠1 = 50°, then m∠3 = ■.
★10. If m∠5 + m∠4 = 140°, then m∠6 = ■.
★11. If m∠4 + m∠5 = 135°, then m∠4 = ■.
★12. If m∠4 = x°, then m∠1 = ■.
★13. If m∠1 = y°, its supplement measures ■.

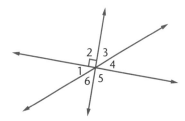

Problem SOLVING

Find the degree measure of the third angle of each triangle.

14.

15.

16.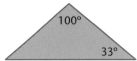

17. Draw vertical angles that measure 55°. What is the measure of the other pair of vertical angles?

18. Angles A and B are supplementary. If m∠A = 60° and m∠B = 4x, what is the value of x?

19. Can two supplementary angles have the same measure? Explain.

20. The measure of an angle is twice the measure of its complement. What is the measure of the angle?

6.1 Angle Relationships 145

6.2 Parallel Lines and Angles

Objective: to find the measures of angles formed by parallel lines; to identify parallel lines

The Traverse City Planning Commission is planning to have the curbs in the business district replaced. The construction company wants to know the angles of the corners ahead of time, so they can build the molds that hold the cement.

When two parallel streets are intersected by another street, eight corners or angles are formed. Is there a method the contractors can use to find the measures of all the angles without having to measure each one?

Two **lines** in the same plane that do not intersect are called **parallel lines**. A line that intersects two or more other lines at different points is a **transversal**. Some pairs of angles formed by parallel lines and a transversal have special names.

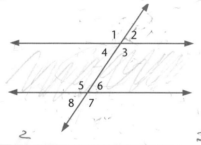

Corresponding angles lie on the same side of the transversal in similar positions.	**Alternate interior angles** lie between the two lines on opposite sides of the transversal.	**Alternate exterior angles** lie outside the two lines and on opposite sides of the transversal.
Angles 2 and 6 are corresponding. Can you name another pair of corresponding angles?	Angles 3 and 5 are alternate interior angles. Can you name another pair of alternate interior angles?	Angles 1 and 7 are alternate exterior angles. Can you name another pair of alternate exterior angles?

When a transversal intersects two parallel lines, the following statements are true:
- Corresponding angles are congruent.
- Alternate interior angles are congruent.
- Alternate exterior angles are congruent.

The reversal is also true. If the corresponding angles, alternate interior angles, or alternate exterior angles are congruent, the lines are parallel.

Example

In the diagram below, line a is parallel to line b and m∠2 = 110°. Find m∠5 and m∠6.

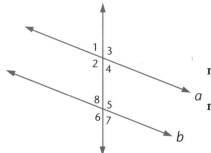

m∠5 = m∠2 = 110°   Alternate interior angles are congruent.

m∠6 = m∠2 = 110°   Corresponding angles are congruent.

Try THESE

Identify each pair of angles as *corresponding, alternate interior, alternate exterior,* or *other*.

1. ∠1 and ∠5
2. ∠1 and ∠6
3. ∠2 and ∠5
4. ∠7 and ∠8
5. ∠3 and ∠8
6. ∠4 and ∠7
7. ∠3 and ∠6
8. ∠2 and ∠6

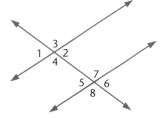

Exercises

In the diagram, lines *m* and *n* are parallel. If the m∠2 = 150°, find the measure of each angle.

1. ∠4
2. ∠6
3. ∠8
4. ∠7
5. ∠1
6. ∠3

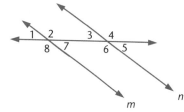

In each diagram, lines *a* and *b* are parallel. Find the measures of the numbered angles.

7.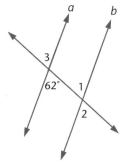

8.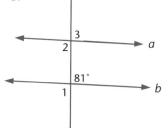

9.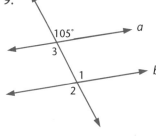

6.2 Parallel Lines and Angles   147

Determine which pairs of lines, if any, are parallel. Explain.

10.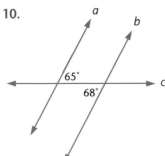

11.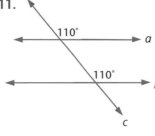

12.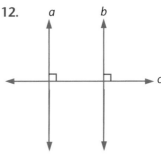

Problem SOLVING

13. If two lines are intersected by a transversal and the corresponding angles are not equal, are the two lines parallel? Explain.

Constructed RESPONSE

★14. If two parallel lines are intersected by a transversal, how are the interior angles on the same side of the transversal related? Draw a diagram, and explain.

Test PREP

15. What is the measure of an angle that is supplementary to a 75° angle?

   a. 25° b. 105° c. 15° d. 30°

Mind BUILDER

Angles

Use the diagram to solve the following problems.

1. Find the measures of ∠1, ∠2, ∠3, ∠4, ∠5, and ∠6.
2. Find the measures of ∠7, ∠8, ∠9, ∠10, ∠11, ∠12, ∠13, and ∠14.
3. Name four angles that are supplementary to ∠2.
4. Under what conditions could ∠1 have the same measure as ∠4?

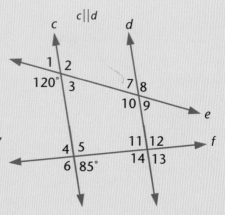

148   6.2 Parallel Lines and Angles

Cumulative Review

Write in standard form.

1. 1.3^2
2. 6^3
3. $\sqrt{0.81}$
4. 30^2
5. $\sqrt{121}$
6. $(4 \cdot 10^2) + (3 \cdot 10^1) + (8 \cdot 10^0)$
7. $(8 \cdot 10{,}000) + (6 \cdot 100) + (2 \cdot 10)$

Estimate.

8. $3.21 + 6.84$
9. $483 - 78$
10. 72×39
11. $73\overline{)296}$
12. $0.18\overline{)0.615}$

Copy and complete each sequence.

13. 1, 6, _____, 216, _____, _____
14. 25, 31, _____, 43, _____, _____, 61
15. 4, _____, 16, 25, _____, _____
16. 4, 8, _____, 32, _____, _____

Solve each equation. Check your solution.

17. $a - 2 = 0$
18. $3 + c = 3.5$
19. $d - 9 = 13$
20. $5e = 45$
21. $3.1 = d - 6.6$
22. $\frac{h}{7} = 2.1$
23. $1.2 = \frac{m}{2}$
24. $5m - 2 = 68$
25. $2 + 3b = 18.86$

Translate each of the following into an equation. Then solve each equation.

26. Andy skied down a slope in 72.3 seconds. This was 4.5 seconds slower than his best time. What was his best time?

27. The width of a gymnasium is 0.6 of the length. The gymnasium is 60 meters long. How wide is it?

Solve.

28. Tony drives 651.3 miles. Elizabeth drives 293.6 miles farther than Tony. How many miles does Elizabeth drive?

29. Alicia averages 79.5 strokes for four rounds of golf. What is her total number of strokes?

30. Chris Messer orders a truckload of sand. She uses 0.5 of it in the patio, 0.33 in the sidewalk, and 0.167 in a sandbox. How much is left?

6.3 Constructions: Line Segments and Angles

Objective: to construct line segments and angles

You can use a compass and a straightedge to copy a line and an angle.

Constructions

A. Construct a **line segment** congruent to a given line segment.

Given: $\overline{AB}$

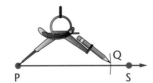

1. Use a straightedge to draw $\overrightarrow{PS}$.

2. Open the compass to match $\overline{AB}$.

3. Keep the compass at the same setting and use P as the center. Draw an arc that intersects $\overrightarrow{PS}$ at Q.

   $\overline{PQ} \cong \overline{AB}$ Line segment PQ is congruent ($\cong$) to line segment AB.

B. Construct an angle congruent to a given angle.

Given: $\angle ABC$

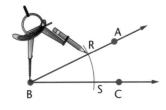

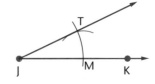

1. Use a straightedge to draw $\overrightarrow{JK}$.

2. Using B as the center, draw an arc that intersects $\angle ABC$ at R and S.

3. Keep the compass at the same setting and use J as the center. Draw an arc that intersects $\overrightarrow{JK}$ at M.

4. Open the compass to match $\overline{RS}$. Draw an arc using M as the center. Label the intersection of the arcs T. Draw $\overrightarrow{JT}$.

   $\angle TJM \cong \angle ABC$

Try THESE

Use a compass and a straightedge to complete the following constructions.

1. Trace $\overline{AB}$ and perform Construction A.

2. Trace $\angle CAB$ and perform Construction B.

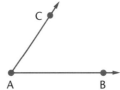

Exercises

Draw two segments, $\overline{XY}$ and $\overline{PT}$, with $\overline{XY} > \overline{PT}$.

1. Construct a segment whose length is the sum of $\overline{XY}$ and $\overline{PT}$.
2. Construct a segment whose length is the difference between $\overline{XY}$ and $\overline{PT}$.

Draw an acute angle WXY and an obtuse angle CAD.

3. Construct $\angle TRS$ congruent to $\angle CAD$.
4. Construct $\angle FGH$ congruent to $\angle WXY$.

Problem SOLVING

★5. Draw a line segment. Then construct an equilateral triangle with the line segment as one side.

MiXeD REVIEW

Write in expanded form using 10^0, 10^1, 10^2, and so on.
For example: $2{,}761 = (2 \cdot 10^3) + (7 \cdot 10^2) + (6 \cdot 10^1) + (1 \cdot 10^0)$

6. 245
7. 1,587
8. 1,040
9. 12,074
10. 70,475
11. 153,000
12. 1,004,578
13. 2,506,000,000

6.4 Constructions: Bisectors of Line Segments and Angles

Objective: to bisect line segments and angles

An architect uses a compass and a straightedge to draw a design for a new building.

You can use a compass and a straightedge to **bisect,** or cut into two congruent parts, line segments and angles.

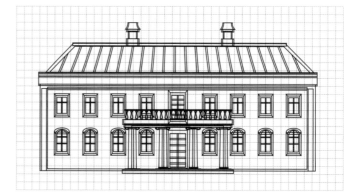

Constructions

A. Bisect a given line segment.

Given: $\overline{XY}$

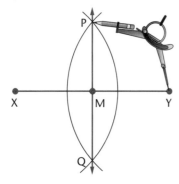

Procedure

1. Open the compass to more than half the length of $\overline{XY}$.
2. Draw two arcs using X and Y as the centers.
3. Label the intersections of the arcs P and Q.
4. Draw $\overleftrightarrow{PQ}$.
5. $\overline{XY}$ is bisected at the midpoint M.

    $\overline{XM} \cong \overline{MY}$

B. Bisect a given angle.

Given: $\angle NAM$

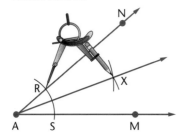

Procedure

1. Using A as the center, draw an arc that intersects $\angle NAM$ at R and S.
2. Draw two arcs using R and S as the centers.
3. Label their intersection X.
4. Draw $\overrightarrow{AX}$. This ray bisects $\angle NAM$.

    $\angle NAX \cong \angle XAM$

Try THESE

Use a compass and a straightedge to complete the following constructions.

1. Trace $\overline{XY}$ and perform Construction A.

2. Trace $\angle NAM$ and perform Construction B.

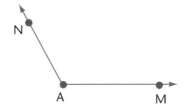

Exercises

Trace each of the following from the figure shown at the right. Then construct the bisector of each figure.

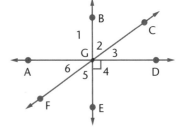

1. $\overline{AD}$ 2. $\angle AGF$ 3. $\overline{BE}$ 4. $\angle DGF$

Trace each angle shown at the right. Then construct angles that have each of the following degree measures.

5. $\frac{1}{2}x$ 6. $\frac{1}{2}y$ 7. $\frac{1}{4}x$

Problem SOLVING

8. Show how a line can bisect a line segment without being perpendicular.

★9. Draw a straight angle. Then bisect it. What type of angles are formed?

10. In what quadrilaterals will the angle bisectors form two diagonals?

MiXeD REVIEW

Write each fraction in simplest form.

11. $\frac{15}{25}$ 12. $\frac{6}{20}$ 13. $\frac{9}{36}$ 14. $\frac{12}{42}$ 15. $\frac{18}{45}$ 16. $\frac{88}{100}$

6.5 Constructions: Perpendiculars and Parallels

Objective: to construct parallel and perpendicular lines

Perpendicular lines are two lines that intersect to form right angles.

Parallel lines are lines in the same plane that do not intersect.

Constructions

A. At a given point on the line, construct a perpendicular line.

Given:

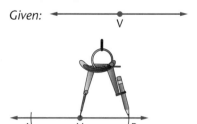 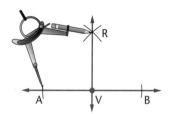

1. Using V as the center, draw two arcs to intersect the line. Label the intersections A and B.

2. Draw two arcs using A and B as the centers. Label their intersection R. Draw $\overleftrightarrow{RV}$.

$\overleftrightarrow{RV} \perp \overleftrightarrow{AB}$ at point V.

B. Construct a line perpendicular to the line through the given point not on the line.

Given:

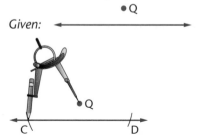 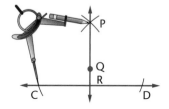

1. Using Q as the center, draw two arcs to intersect the line. Label the intersections C and D.

2. Draw two arcs using C and D as the centers. Label their intersection P. Draw $\overleftrightarrow{PQ}$.

$\overleftrightarrow{PR} \perp \overleftrightarrow{CD}$ through point Q at point R.

C. Construct a line parallel to the given line.

Given: $\overleftrightarrow{AB}$

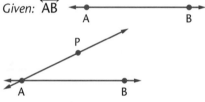 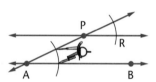

1. Place point P above the line. Draw a line through P to intersect $\overleftrightarrow{AB}$ at A.

2. With P as a vertex, construct an angle congruent to ∠PAB. Draw $\overleftrightarrow{PR}$.

$\overleftrightarrow{PR} \parallel \overleftrightarrow{AB}$

Try THESE

Trace the given figure to complete each of the following constructions.

1. Construct a line perpendicular to the line at P.

2. Construct a line through S perpendicular to $\overleftrightarrow{QR}$.

3. Construct a line through J parallel to $\overleftrightarrow{GH}$.

Exercises

Trace the figure shown at the right to complete Exercises 1–3.

1. Construct a line through N perpendicular to $\overleftrightarrow{KL}$.
2. Construct a line through M parallel to $\overleftrightarrow{KL}$.
3. Construct a line through L parallel to $\overleftrightarrow{MN}$.
4. Construct $\overleftrightarrow{AB} \perp \overleftrightarrow{CD}$. Label the point of intersection as X.
5. Using the construction in Exercise 4, construct $\overleftrightarrow{EF} \perp \overleftrightarrow{AB}$ at point A.
6. State the relationship between $\overleftrightarrow{EF}$ and $\overleftrightarrow{CD}$ in Exercise 5.

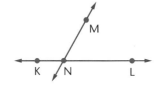

Problem SOLVING

7. Trace $\overline{AB}$. Construct the perpendicular bisector of $\overline{AB}$.

8. Trace $\overline{EF}$. Construct a square with $\overline{EF}$ as one of the sides.

★9. Trace $\overline{CD}$. Construct a right triangle with $\overline{CD}$ as one of the legs.

10. $\overrightarrow{XS}$ and $\overrightarrow{YT}$ are perpendicular to $\overleftrightarrow{XY}$. How do $\overrightarrow{XS}$ and $\overrightarrow{YT}$ relate to each other?

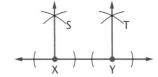

6.5 Constructions: Perpendiculars and Parallels

6.6 Problem-Solving Application: Constructing Golden Rectangles

Objective: to construct a golden rectangle; to solve problems regarding golden rectangles

The ancient Greeks often used a certain kind of rectangle, known as the **golden rectangle**, in their architecture and art. The golden rectangle has the special property that the length divided by the width is approximately 1.6. The front of the famous Parthenon has the shape of a golden rectangle.

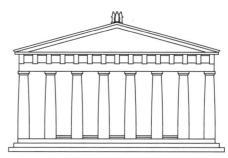

You can use a compass and a straightedge to construct a golden rectangle.

1. Construct a square.

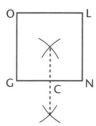

2. Bisect $\overline{GN}$. Label the midpoint C.

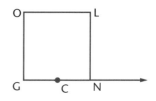

3. Extend side GN as shown.

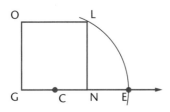

4. Using C as the center, draw an arc through L that intersects $\overrightarrow{GN}$ at E.

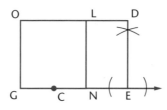

5. Construct a line perpendicular to $\overrightarrow{GN}$ at E. Extend side OL to intersect the perpendicular line at D.

> The measure of $\overline{GE}$ divided by the measure of $\overline{OG}$ is about 1.6. This figure is a golden rectangle.

Try THESE

Use the given measures to determine if rectangle ABCD is a golden rectangle. Write *yes* or *no*.

1. AB = 14.56, AD = 9
2. BC = 25, CD = 40.45
3. AD = 8.5, DC = 13
4. BC = 6.8, AB = 11

156   6.6 Problem-Solving Application: Constructing Golden Rectangles

Solve

1. Construct a square measuring 4 cm on each side. Then construct a golden rectangle.

2. Is rectangle LDEN, as seen in part 5 of the Construction on p. 156, a golden rectangle? Measure and divide to determine your answer.

3. The Great Math Detective likes to amaze his friends by performing math tricks. Here is one for you.
   - He invited four friends to share a large, square pizza.
   - He made four straight cuts that went from edge to edge of the pizza.
   - Everyone at the party had two pieces of pizza.
   - How did the Great Math Detective cut the pizza?

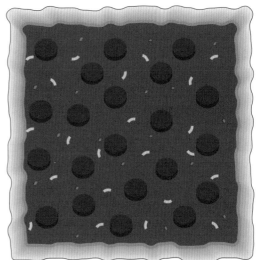

Mid-Chapter REVIEW

Use the figure at the right to solve the following problems.

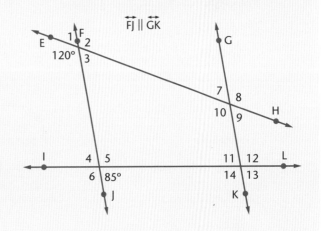

1. Find the measure of ∠1.
2. Find the measure of ∠9.
3. Find the measure of ∠11.
4. If m∠7 = 60°, then m∠8 = ■.
5. If m∠11 = 80°, then m∠12 = ■.
6. Name a pair of angles that are complementary.
7. Name an angle that is supplementary to ∠2.
8. Name a pair of alternate interior angles.
9. Name a pair of alternate exterior angles.
10. Name a pair of corresponding angles.
11. Under what conditions could ∠8 have the same measure as ∠5?

6.6 Problem-Solving Application: Constructing Golden Rectangles

6.7 Triangles and Quadrilaterals

Objective: to classify triangles and quadrilaterals

Triangle braces are used in the construction of roller coasters. **Triangles** can be classified by the number of congruent sides.

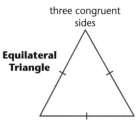
Equilateral Triangle — three congruent sides

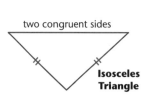
Isosceles Triangle — two congruent sides

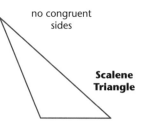
Scalene Triangle — no congruent sides

More Examples

A. Triangles can also be classified by angles. All triangles have at least two **acute angles** (between 0°–90°); therefore, the third angle is used to classify the triangle.

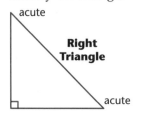
Right Triangle — acute, acute

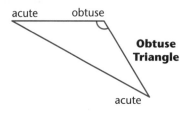
Obtuse Triangle — acute, obtuse, acute

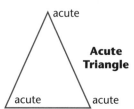
Acute Triangle — acute, acute, acute

B. Quadrilaterals are classified as follows.

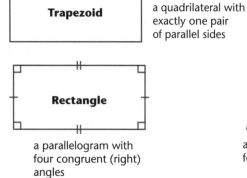

Trapezoid — a quadrilateral with exactly one pair of parallel sides

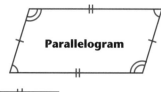
Parallelogram — a quadrilateral with two pairs of parallel sides

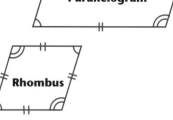
Rhombus — a parallelogram with four congruent sides

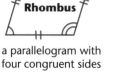

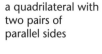
Square — a parallelogram with four congruent sides and four congruent (right) angles

a parallelogram with four congruent (right) angles (Rectangle)

Try THESE

Classify each triangle by its sides and then by its angles.

1.
2.
3.
4.

Exercises

Draw a triangle that satisfies both of the classifications. If no such triangle exists, write *none*.

1. isosceles, right
2. equilateral, obtuse
3. scalene, acute

List all quadrilaterals with the following characteristics.

4. two pairs of parallel sides
5. exactly one pair of parallel sides
6. all sides congruent
7. all angles congruent

State whether each of the following is *always true*, *sometimes true*, or *never true*.

8. A rectangle is a parallelogram.
9. A rectangle is a square.
10. A rhombus is a square.
11. A trapezoid is a parallelogram.
12. A square is a parallelogram.
★13. A square is a rhombus.
14. Isosceles triangles are equilateral.
15. Equilateral triangles are isosceles.

Problem SOLVING

16. Show that a parallelogram can be formed from two congruent trapezoids. Then show a different parallelogram.

17. Show that a kite shape can be divided into two non-congruent isosceles triangles. Show that it can also be divided into two congruent scalene triangles.

18. Draw a Venn diagram showing the relationship of quadrilaterals.

Constructed RESPONSE

★19. Cut six congruent equilateral triangles from a sheet of paper. Piece them together to form a hexagon. Why do the triangles form a perfect fit? What other regular features can be made using at least two of the triangles?

20. Does a kite shape fit any of the classes of quadrilaterals shown on p. 158? Explain.

21. Draw a general quadrilateral (no sides congruent or parallel). What is the sum of the measures of the angles? Is this sum the same for any quadrilateral? Show why or why not.

Mixed REVIEW

Solve each equation.

22. $z + 7 = 35$
23. $52 = s - 14$
24. $\dfrac{k}{4} = 52$
25. $20.8 = 16t$
26. $1.7 = \dfrac{n}{13}$
27. $27 = 2c + 3$
28. $\dfrac{p}{7} - 4 = 35$
29. $23 = \dfrac{r}{12} + 15$

6.8 Polygons and Angles

Objective: to find angle measures in a polygon

Polygons are closed figures in a plane formed by line segments. A polygon in which all sides have the same length and all angles have the same degree measure is called a **regular polygon**. To the right is a list of polygons.

Polygon	Number of Sides	Polygon	Number of Sides
Triangle	3	Octagon	8
Quadrilateral	4	Nonagon	9
Pentagon	5	Decagon	10
Hexagon	6	Dodecagon	11
Heptagon	7		

You know that the sum of the measure of the angles of a triangle is 180°. A quadrilateral is made up of two triangles. The sum of its angles is 360°. The sum of the angles in a polygon depends on the number of sides of the polygon.

> **Polygon Angle Measures**
>
> The sum of the angle measures of a polygon with n sides is $(n - 2) \cdot 180$.
>
> The measure of one angle in a regular polygon with n sides is $\frac{(n - 2) \cdot 180}{n}$.

Examples

A. Find the sum of the angle measures of an octagon.

An octagon has 8 sides.

$(n - 2) \cdot 180 = (8 - 2) \cdot 180$ Replace n with 8.

$\qquad\qquad\quad = 1{,}080°$ Simplify.

B. Find the missing angle measure in the hexagon below.

Find the sum of the angle measures of a hexagon.

$(n - 2) \cdot 180 = (6 - 2) \cdot 180$ Replace n with 6.

$\qquad\qquad\quad = 720°$

Write an equation. Let $x =$ the missing angle measure.

$720° = 130° + 110° + 110° + 130° + 120° + x$

$720° = 600° + x$

$120° = x$

The missing angle is 160°.

Try THESE

Find the sum of the angle measures of the interior angles of each polygon.

1. pentagon
2. heptagon
3. nonagon
4. decagon

Exercises

Find the value of x in each polygon.

1. (pentagon with angles 140°, 105°, 85°, 110°, x)
2. (heptagon with angles 150°, 104°, 150°, x, 145°, 147°, 118°, 127°)
3. (hexagon with angles 136°, 136°, 128°, 115°, x, 115°, 128°)
4. (arrow shape with angles x, 60°, x)
5. (rectangle with angles x, x, x, x)
6. (hexagon with angles x, 2x, 2x, x, 2x, 2x)

Find the measure of each angle of a regular polygon with the given number of sides.

7. 4 sides
8. 5 sides
9. 12 sides

Problem SOLVING

10. Four of the angle measures of a pentagon are 85°, 110°, 135°, and 95°. Find the measure of the missing angle.
11. A geodesic dome has some panels that are regular hexagons. Draw a regular hexagon. Find the measure of each angle in the regular hexagon.
★12. The sum of the measures of a polygon is 1,620°. How many sides does it have?

Test PREP

13. The sum of the measures of the interior angles of an octagon is ____.
    a. 1,080°
    b. 1,620°
    c. 900°
    d. 1,260°

14. The measure of one angle in a regular pentagon is ____.
    a. 180°
    b. 90°
    c. 108°
    d. 120°

6.9 Congruent Figures

Objective: to identify congruent triangles

The cells in a bee's honeycomb have the same size and shape. Figures that have the same size and shape are **congruent**.

In general, two geometric figures are congruent ($\cong$) if one of them can be placed on top of the other and fit exactly, point for point, side for side, and angle for angle.

If the corresponding sides and angles of two triangles are congruent, the triangles are congruent.

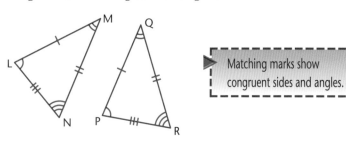

Matching marks show congruent sides and angles.

Say: △LMN is congruent to △PQR.

Write: △LMN $\cong$ △PQR

More Examples

You can determine that two triangles are congruent without knowing the measures of all sides and angles.

A. Side-Side-Side (**SSS**)
When the corresponding sides of two triangles are congruent, the triangles are congruent.

$\overline{AB} \cong \overline{DE}$
$\overline{BC} \cong \overline{EF}$
$\overline{AC} \cong \overline{DF}$

B. Side-Angle-Side (**SAS**)
When two corresponding sides and the included angle are congruent, the triangles are congruent.

$\overline{LM} \cong \overline{QR}$
∠MLN $\cong$ ∠RQS
$\overline{LN} \cong \overline{QS}$

C. Angle-Side-Angle (**ASA**)
When two corresponding angles and the included side are congruent, the triangles are congruent.

∠YXZ $\cong$ ∠HGJ
$\overline{XZ} \cong \overline{GJ}$
∠XZY $\cong$ ∠GJH

Exercises

Use the triangles at the right to solve the following problems.

1. Name one pair of congruent angles.
2. Name two pairs of congruent sides.
3. Are the triangles congruent by **SSS**, **SAS**, or **ASA**?

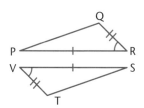

Use the congruent triangles shown at the right to complete each of the following.

4. $\overline{JK} \cong$ ■   5. $\angle JKL \cong$ ■   6. $\angle KJL \cong$ ■

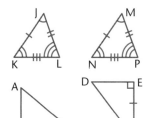

7. $\angle ABC \cong$ ■   8. $\angle ACB \cong$ ■   9. $\overline{AB} \cong$ ■

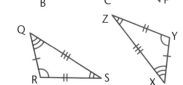

10. $\angle QRS \cong$ ■   11. $\overline{QR} \cong$ ■   12. $\overline{RS} \cong$ ■

Is each pair of triangles congruent by SSS, SAS, or ASA?

13. 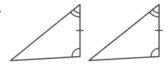   14. 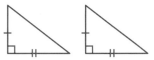   15.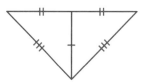

16. 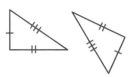   17. 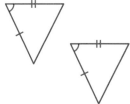   18.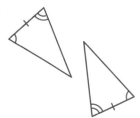

6.9 Congruent Figures

6.10 Translations and Rotations

Objective: to translate and rotate figures

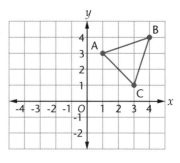

Consider the triangle shown on the coordinate system at the right. It has vertices at points A(1, 3), B(4, 4), and C(3, 1).

Examples

A. Suppose you switch the coordinates and multiply the new *x*-coordinate by -1.

A(1, 3) → A'(-3, 1)
B(4, 4) → B'(-4, 4)
C(3, 1) → C'(-1, 3)

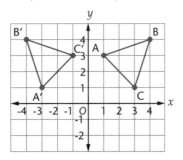

△A'B'C' is a rotation of △ABC counterclockwise 90° around the origin.

B. Suppose you add -3 to the *y*-coordinate of each point.

A(1, 3) → A'(1, 0)
B(4, 4) → B'(4, 1)
C(3, 1) → C'(3, -2)

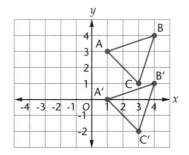

△A'B'C' is a translation of △ABC down 3 units.

Rotations

A **rotation** is a transformation that turns each point the same number of degrees around a common point. You will rotate around the origin.

To rotate a point **90°** counterclockwise, switch the coordinates, and then multiply the new *x*-coordinate by -1.
$$P(2, 6) \rightarrow P'(-6, 2) \qquad P(x, y) \rightarrow P'(-y, x)$$

To rotate a point **180°** counterclockwise, multiply the coordinates by -1.
$$P(2, 6) \rightarrow P'(-2, -6) \qquad P(x, y) \rightarrow P'(-x, -y)$$

Translations

A **translation** is a transformation that moves each point on a figure the same distance and in the same direction.

Slide to the **left** *a* units: $x \rightarrow x - a$

Slide to the **right** *a* units: $x \rightarrow x + a$

Slide **up** *b* units: $y \rightarrow y + b$

Slide **down** *b* units: $y \rightarrow y - b$

Complete the following.

1. Multiply the *x*- and *y*-coordinates by -1 of (3, 4).
2. Add -2 to the *x*- and *y*-coordinates of (-6, -11).
3. Add 4 to the *x*-coordinate of (-7, 5).

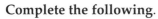

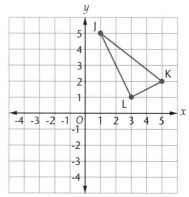

Copy △JKL on graph paper. Then draw the triangle that results after completing each of the following.

1. Translate △JKL down 3 units.
2. Translate △JKL left 2 units and up 2 units.
3. Rotate △JKL 90°.
4. Rotate △JKL 180°.

Complete the following.

5. On graph paper, graph the points P(-2, 4), Q(1, 3), and R(-1, -2). Translate each point 2 units up. Write the coordinates of each new point, and graph the new figure.

6. On graph paper, graph the points A(1, 6), B(2, 2), and C(5, 4). Rotate each point 90°. Write the coordinates of each new point, and graph the new figure.

7. On graph paper, graph the points K(-3, 3), L(-1, 3), M(0, 1), and N(-4, 1). Translate each point 5 units right. Then rotate each point 180°. Write the coordinates of each new point, and graph the new figure.

8. On graph paper, graph the points W(-4, 3), X(1, 2), Y(2, -2), and Z(-3, -1). Rotate the figure 90°, and graph the new figure. Then add 2 to the *y*-coordinate of each point. Write the coordinates of each new point, and graph the new figure.

Problem SOLVING

9. On a coordinate system, identify 5 points that form the vertices of a pentagon. Translate the pentagon 3 units down and 1 unit left. Write the coordinates of the new figure, and graph the new figure.

Constructed RESPONSE

10. On graph paper, graph the points M(0, 0), N(0, 2), O(3, 2), and P(3, 0).
    a. Rotate the figure 90° two times.
    b. Write the coordinates of each new point, and graph each new figure.
    c. Then rotate the figure 90° two more times. Write the coordinates of each new point.
    d. What do you notice about the coordinates of the original figure and the new figure? Explain.

MiXeD REVIEW

Solve.

11. A sweater used to sell for $45.00. It now sells for $38.25. What is the percent of decrease?

12. What is the simple interest to the nearest cent if you borrow $1,500 at 11% interest for 2 years?

Mind BUILDER

Logical Thinking I

Trace the figure shown at the right. Color 3 squares blue, 3 yellow, and 3 red. No two squares with a common boundary may be the same color.

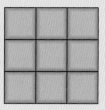

Problem Solving

Tricky Cuts

Each shape below can be cut into two identical shapes by making only one cut. The cut does not have to be straight. An example is shown at the right.

Trace each shape below, and draw a line or make a cut so that the resulting shapes are identical.

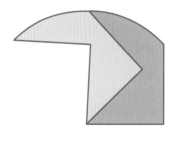

Extension

Draw a figure that has no straight edges and can be cut into identical shapes.

6.11 Reflections and Symmetry

Objective: to reflect figures and identify lines of symmetry

The mirror image produced by flipping a figure over a line is called a **reflection**. The line the figure is flipped over is called the line of reflection.

Examples

A. Suppose you multiply the x-coordinate of each point by -1.

   A(1, 3) → A'(-1, 3)
   B(4, 4) → B'(-4, 4)
   C(3, 1) → C'(-3, 1)

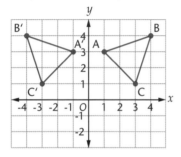

△A'B'C' is a reflection of △ABC about the y-axis.

B. Suppose you multiply the y-coordinate of each point by -1.

   A(1, 3) → A'(1, -3)
   B(4, 4) → B'(4, -4)
   C(3, 1) → C'(3, -1)

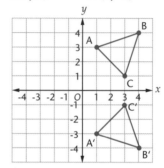

△A'B'C' is a reflection of △ABC about the x-axis.

Reflections

To reflect a point about the **x-axis**, multiply its y-coordinate by -1.
P(x, y) → P'(x, -y)

To reflect a point about the **y-axis**, multiply its x-coordinate by -1.
P(x, y) → P'(-x, y)

A line can be drawn through some figures so that the figure on one side of the line is a mirror image of the figure on the other side. This line is called a **line of symmetry**.

More Examples
How many lines of symmetry does each figure have?

C.

It has 1 line of symmetry.

D.

It has 2 lines of symmetry.

E.

It has no lines of symmetry.

Try THESE

Write the new coordinates.

1. Reflect the point (5, 8) about the x-axis.
2. Reflect the point (-6, 2) about the y-axis.
3. Reflect the point (-3, -4) about the x-axis.

Exercises

Copy △JKL on graph paper. Then draw the triangle that results after completing problems 1–4.

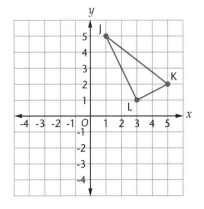

1. Reflect △JKL about the x-axis.
2. Reflect △JKL about the y-axis.
3. Reflect △JKL about the x-axis and then about the y-axis.
4. Reflect △JKL about the y-axis and then about the x-axis.
5. On graph paper, graph the points A(1, 6), B(2, 2), and C(5, 4). Reflect each point about the x-axis. Write the coordinates of each new point, and graph the new figure.
6. On graph paper, graph the points G(-3, -2), H(-4, -4), and I(-1, -2). Reflect each point about the x-axis. Then reflect the points about the y-axis. Write the coordinates of each new point, and graph the new figure.
7. On graph paper, graph the points K(1, -2), L(3, -2), M(4, -4), and N(0, -4). Translate each point 5 units up, and then reflect the figure about the y-axis. Write the coordinates of each new point, and graph the new figure.

How many lines of symmetry does each figure have? Draw all lines of symmetry. If it does not have any, state *none*.

8.

9.

10.

6.11 Reflections and Symmetry

6.12 Problem-Solving Strategy: Identifying Necessary Facts

Objective: to identify necessary facts to solve word problems

Mr. Reinhardt planted $\frac{5}{8}$ of his farm in corn, $\frac{1}{6}$ in soybeans, and $\frac{5}{24}$ in wheat. How many acres are planted in corn and soybeans?

Sometimes a problem may have more facts than you need to solve it. You must carefully choose the facts necessary to solve the problem.

A problem may not have all the facts you need to solve it. Carefully study the facts in the problem above.

 You need to find the number of acres planted in corn and soybeans. You know the fraction of the farm planted in each.

 To find the number of acres planted in corn and soybeans, you must perform two steps. First add the fractions planted in corn and soybeans. Then multiply that fraction by the number of acres on the farm.

Corn plus soybeans is part planted in corn and soybeans.

$$\frac{5}{8} + \frac{1}{6} = \frac{15}{24} + \frac{4}{24} = \frac{19}{24}$$

$\frac{5}{24}$ planted in wheat is an extra fact.

$\frac{19}{24}$ × the number of acres = acres planted in corn and soybeans

You cannot multiply unless you know the number of acres on the farm.

 The problem does not have all the facts you need to solve it.

State any extra or missing facts. Write *can be solved* or *cannot be solved* for each problem.

1. A $6\frac{1}{2}$-foot board is to be cut into pieces of the same length. How long is each piece?

2. Mr. White buys $5\frac{1}{2}$ pounds of cookies. How much change does he receive from a ten-dollar bill?

Solve

Solve if possible. State any extra facts or any missing facts.

1. Jane has a test average of 82% in science. She correctly answered $\frac{3}{4}$ of the problems on her last test. If the test had 20 problems, how many did she answer correctly?

2. Jerry uses $1\frac{1}{3}$ square yards of material to make a wall hanging. At $3.00 a square yard, what is the cost of the material for 3 wall hangings?

3. Joan Alioto earns $4.50 an hour as a fitness advisor. She works 20 hours a week. In how many hours will she earn $180?

4. A car with a 4-cylinder engine uses $8\frac{1}{2}$ gallons of gasoline on a 5-hour trip. How many miles per gallon is this?

5. In 5 years, the stock in an athletic shoe company went up $11.50 to a price of $20.25 a share. What is the average increase each year?

6. A bus gets 6.5 miles per gallon. The bus is driven about 220 miles a day. How many miles can the bus go on 22.6 gallons of gasoline?

7. A recipe calls for $2\frac{2}{3}$ cups of flour. Charlene has to triple the recipe. How many cups of flour does she need?

8. The population of Detroit is 2.36 times the population of Boston. What is the population of Detroit?

9. The six employees at Harris Sports Store work 10, 15, 12, 20, 40, and 40 hours each week. To the nearest tenth of an hour, what is the average weekly working time?

10. The odometer reading is 4,206.9 miles at the start of a trip and 4,365 at the end. Find the miles per gallon if 6.2 gallons of gasoline are used.

11. A school buys 3 grosses of pencils. Each gross (144 pencils) costs $10.80. Find the cost of each pencil.

12. Rosalind Lucero paid $1,536.00 tax in 1999 and $1,623.60 in 2000. Find the average tax she paid each month in 1999.

13. Eight adult tickets to a game cost $14.00. Children's tickets cost $0.75 each. What is the cost of each adult ticket?

14. Mr. Cardenas paid $50.00 down on a set of golf clubs and made 12 payments of $21.50. Find the total amount he paid for the clubs.

6.12 Problem-Solving Strategy: Identifying Necessary Facts

Chapter 6 Review

Language and Concepts

Choose a term from the list at the right to correctly complete each sentence.

1. An angle that measures between 0° and 90° is a(n) _____ angle.
2. Two angles are _____ if the sum of their measures equals 90°.
3. When a(n) _____ intersects two parallel lines, the corresponding angles are congruent.
4. A(n) _____ triangle is a triangle with three congruent sides.
5. A(n) _____ triangle is a triangle that has two congruent sides.
6. An angle that measures between 90° and 180° is a(n) _____ angle.
7. _____ angles are formed by two intersecting lines.
8. A(n) _____ triangle is a triangle that has no congruent sides.
9. A parallelogram with four congruent sides is a(n) _____.
10. Two angles are _____ if the sum of their measures equals 180°.

equilateral
isosceles
scalene
obtuse
acute
vertical
trapezoid
rhombus
right
transversal
complementary
supplementary

Skills and Problem Solving

Use the figure at the right to complete each of the following. Section 6.1

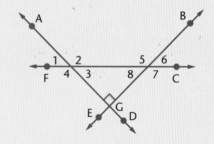

11. ∠3 and ■ are complementary.
12. ∠3 and ■ are supplementary.
13. Name an angle adjacent to ∠5.
14. ∠6 and ■ are vertical angles.
15. m∠3 + m∠8 + m∠AGB = ■

Use the figure at the right to complete each of the following. $\overleftrightarrow{JL} \parallel \overleftrightarrow{MP}$ Section 6.2

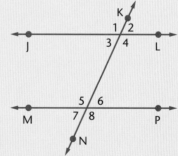

16. Name a pair of corresponding angles.
17. Name a pair of parallel lines.
18. Line ■ is a transversal.
19. Name two pairs of alternate interior angles.
20. If m∠4 = 105°, then m∠8 = ■.

Use the figure at the right to complete the following. Sections 6.3–6.4

21. Construct a segment congruent to $\overline{CD}$ and bisect it.

22. Construct an angle congruent to $\angle BCD$ and bisect it.

Complete each of the following. Section 6.5

23. Draw a line DL and a point R not on the line. Construct a line through R perpendicular to $\overleftrightarrow{DL}$.

24. Draw a line MS and a point L not on the line. Construct a line through L parallel to $\overleftrightarrow{MS}$.

Complete each of the following. Section 6.7

25. Draw an isosceles triangle.

26. Draw a scalene triangle.

27. Name the quadrilaterals that have two pairs of parallel sides.

28. Name the triangles that are classified by angles.

Find the value of x in each figure. Section 6.8

29.

30.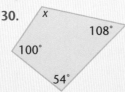

31.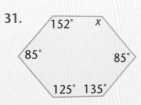

32.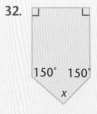

Is each pair of triangles congruent by SSS, SAS, or ASA? Section 6.9

33.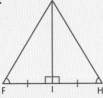

34.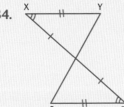

35.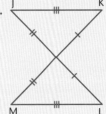

Use the graph at the right to complete the following. Sections 6.10–6.11

36. Translate $\triangle ABC$ 2 units right and 4 units down. Draw the new triangle.

37. Reflect $\triangle ABC$ over the x-axis. Draw the new triangle.

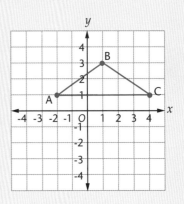

Solve. State any extra or missing facts. Section 6.12

38. Burt spends $\frac{1}{15}$ of his allowance on lunches and $\frac{1}{10}$ on school supplies. He saves $\frac{1}{4}$ of his allowance and spends the rest on miscellaneous items. What part of his allowance does he spend on lunches and school supplies?

Chapter 6 Review   173

Chapter 6 Test

Match the letter of each figure to its most exact description.

1. line segment
2. complementary angles
3. obtuse angle
4. vertical angles
5. supplementary angles
6. perpendicular lines
7. equilateral triangle
8. regular quadrilateral
9. rhombus

a.   b.   c. (line segment)

d.   e.   f.

g. (acute angle figure) h. (perpendicular lines figure) i. (rhombus figure)

Use the figure shown at the right to name each of the following.

10. perpendicular lines
11. right angles
12. vertical angles
13. complementary angles

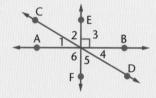

Use the congruent triangles shown at the right to complete each of the following.

14. $\overline{AB} \cong$ ■
15. $\overline{XZ} \cong$ ■
16. $\overline{BC} \cong$ ■
17. $\angle ABC \cong$ ■
18. $\angle XYZ \cong$ ■
19. $\angle YXZ \cong$ ■

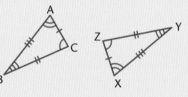

Complete each of the following.

20. Construct a triangle congruent to any triangle using the **SAS** rule.
21. Name the quadrilaterals that have four congruent sides.

Use the graph at the right to complete the following.

22. Rotate △ABC 90° about the origin. Draw the new triangle.
23. Translate △ABC 3 units right. Draw the new triangle.
24. Reflect △ABC over the *y*-axis. Draw the new triangle.

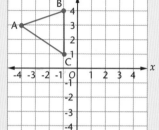

Solve. State any extra or missing facts.

25. Seven pints of blueberries cost $17.50. Three pints of strawberries cost $6.00. Five pints of raspberries cost $24.50. How much do four pints of blueberries and two pints of raspberries cost? Show all work. Explain.

Change of Pace

Tessellations

You can use one or more polygons to form a repetitive pattern. This pattern is called a **tessellation**. In a tessellation, the polygons fit together with no holes or gaps. M.C. Escher (1898–1972) was a Dutch artist who was famous for using tessellations in his art.

In a tessellation, the sum of the angle measures where the vertices meet must equal 360°. You can create a tessellation by repeatedly translating, rotating, or reflecting a figure.

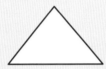 90° + 90° + 90° + 90° = 360°

You can create a tessellation with squares in a plane.

Trace the equilateral triangle shown below. Use the triangle to draw a tessellation.

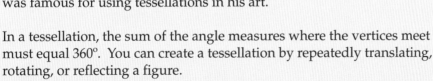

 Rotate, reflect, and translate the equilateral triangle to cover the area.

1. Trace each geometric shape below.
2. Cut out each shape and trace it onto a piece of cardboard.
3. Cut out each shape from the cardboard.
4. Use each cardboard piece to draw a tessellation for each regular polygon.

   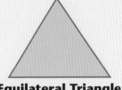

Regular Octagon   **Regular Hexagon**   **Equilateral Triangle**   **Square**

5. Can you make a tessellation from the following combinations?

   a. square and octagon
   b. square, triangle, and hexagon
   c. square, triangle, and dodecagon
   d. another combination of your choice

Change of Pace **175**

Cumulative Test

1.   9.86
     × 0.7
     ---
   a. 6.902
   b. 69.02
   c. 690.2
   d. none of the above

2. $2.88 \div 8 =$ _____
   a. 36
   b. 3.6
   c. 0.36
   d. none of the above

3. Solve.
   $\frac{x}{8} - 3 = 4$
   a. 56
   b. 32
   c. 8
   d. none of the above

4. What is the value of $\sqrt{144} + 2 \cdot (3^2 - 2^3)$?
   a. 14
   b. 56
   c. 126
   d. 118

5. $\$6.07 - \$2.59 =$ _____
   a. $3.48
   b. $3.58
   c. $4.52
   d. $8.66

6. What is the sentence "Four more than y is six" written as an equation?
   a. $4 = y + 6$
   b. $4 + 6 = y$
   c. $4 + y = 6$
   d. none of the above

7. $6(3 + 4) =$ _____
   a. $(6 \cdot 4) - (6 \cdot 3)$
   b. $(6 \cdot 3) + 4$
   c. $(6 \cdot 3) + (6 \cdot 4)$
   d. none of the above

8. Complete the sequence.
   2, 5, 15, 18, 54, ■, ■, ■
   a. 18, 15, 5
   b. 57, 171, 174
   c. 162, 165, 495
   d. none of the above

9. For a home economics project, Silvia buys 2.5 yards of material and 2 patterns. The material costs $3.20 a yard, and the patterns cost $1.95 each. The tax on the material and patterns is $0.48. Which of the following is true?
   a. The two patterns cost $2.80.
   b. The total cost plus tax is $12.38.
   c. The change from $15 is $2.52.
   d. none of the above

10. There were 623 people at the basketball game. There were 458 students and 305 females at the game.

    Which of the following problems can be solved using the information given?
    a. How many students were female?
    b. How many of the people were not students?
    c. How many men were not students?
    d. How many seats were empty?

CHAPTER 7

Measuring Area and Volume

Carolyn Heaton
Japan

7.1 Area

Objective: to find the area of parallelograms, triangles, and trapezoids

An interior decorator uses area to find how much fabric she will need to make items for a house, such as furniture and curtains. The **area** of a figure is the number of square units that cover a surface.

You know that the area of a rectangle is length times width. You can also find the area of many other figures, such as parallelograms, triangles, and trapezoids.

Area of a Parallelogram

The area of a parallelogram is the product of the base b and the height h. The height of a parallelogram is perpendicular to the bases.

$A = bh$

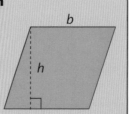

Examples

A. Find the area of the parallelogram.

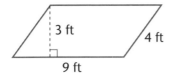

$A = bh$ area of a parallelogram

$A = 9 \cdot 3$ Replace b with 9 and h with 3.

$A = 27$ Multiply.

The area of the parallelogram is 27 ft².   ▶ *Remember: Area is expressed in square units.*

The area of a triangle is related to the area of a parallelogram. A diagonal divides a parallelogram into two congruent triangles. A triangle is one-half of a parallelogram. Using the formula for a parallelogram, you can find the area of a triangle.

Area of a Triangle

The area of a triangle is half the product of the base b times the height h. The height of a triangle is perpendicular to the base.

$A = \frac{1}{2}bh$

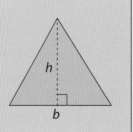

178   7.1 Area

B. Find the area of the triangle.

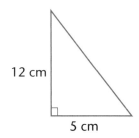

$A = \frac{1}{2}bh$ area of a triangle

$A = \frac{1}{2} \cdot 5 \cdot 12$ Replace b with 5 and h with 12.

$A = 30$ Multiply.

The area of the triangle is 30 cm².

A trapezoid is a quadrilateral with exactly one set of opposite parallel sides. The parallel sides are the bases b_1 and b_2. The height h is a perpendicular line that connects the bases.

Area of a Trapezoid

The area of a trapezoid is half the sum of the bases b_1 and b_2 times the height h.

$A = \frac{1}{2}(b_1 + b_2) \cdot h$

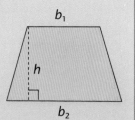

More Examples

C. Find the area of the trapezoid.

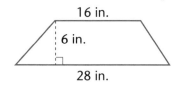

$A = \frac{1}{2}(b_1 + b_2) \cdot h$ area of a trapezoid

$A = \frac{1}{2}(28 + 16) \cdot 6$ Replace b_1 with 28, b_2 with 16, and h with 6.

$A = 132$ in.²   Simplify.

D. Find the area of the irregular figure.

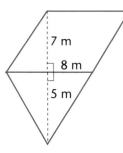

The figure is composed of a parallelogram and a triangle.

$A = bh$ area of a parallelogram

$A = 8 \cdot 7 = 56$ Simplify.

$A = \frac{1}{2}bh$ area of a triangle

$A = \frac{1}{2} \cdot 8 \cdot 5 = 20$ Simplify.

56 m² + 20 m² = 76 m²   You can add the two areas together to find the area of the figure.

The total area of the figure is 76 m².

Try THESE

Find the area of each figure.

1.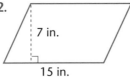

2.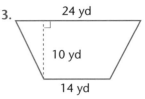

3.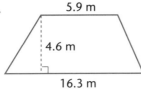

Exercises

Find the area of each figure.

1.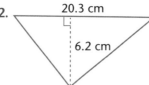

2.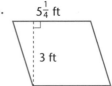

3.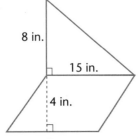

4. triangle: base = 14 cm, height = $5\frac{1}{2}$ cm

5. parallelogram: base = 4.2 m, height = 2.3 m

6. trapezoid: height = 16 in., bases = 12 in. and 21 in.

Find the area of each irregular figure.

7.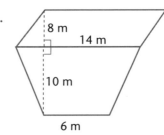

8.

Problem SOLVING

9. Find the height of a triangle with an area of 25 square centimeters and a base of 2.5 centimeters.

10. A trapezoid has an area of 48 square feet. If the lengths of the bases are 16 feet and 8 feet, find the height.

Constructed RESPONSE

11. The height of a triangle is doubled. The length of the base remains the same. How is the area of the triangle affected? Explain, and give an example.

Problem Solving

Carpet the Library

The George Washington School Library is going to be re-carpeted. The dimensions of the room are shown in the floor plan below. All corners are right angles.

How much carpeting is needed if it is bought in square-foot pieces?

How much carpeting must be ordered from a 12-foot-wide roll?

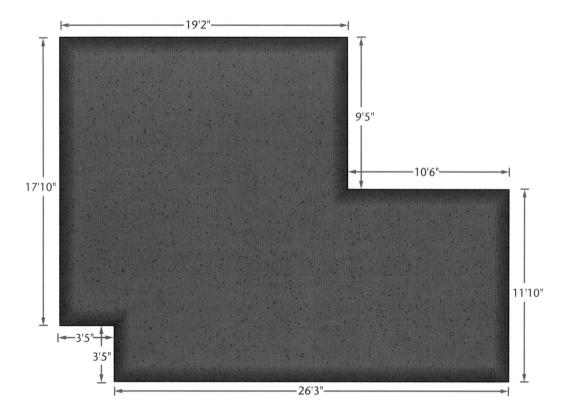

Extension

Suppose the square-foot pieces are priced at $25.00 per dozen, and the carpet on a roll costs $18.50 per square yard. Would you choose to buy the square pieces or the carpet from a roll? Why?

7.2 Circles

Objective: to find the circumference and area of circles

A **circle** is a set of points in a plane that are the same distance from a given point in the plane called the **center**. The line segment that has one endpoint at the center and the other endpoint on the circle is called the **radius** (r). A **diameter** (d) is a line segment that passes through the center with both endpoints on the circle. The diameter is twice the radius. A **chord** is a line segment with both endpoints on the circle.

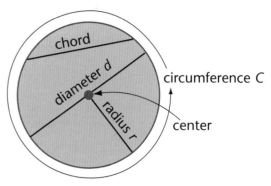

The **circumference** (C) of a circle is the distance around the circle. When finding the circumference of a circle, you need to use **pi**. The Greek letter pi (π) represents the ratio of the circumference of a circle to the diameter of a circle which is always 3.14159 . . . . Approximate values for π are 3.14 and $\frac{22}{7}$.

Circumference of a Circle

The circumference C of a circle is equal to its diameter d times π or 2 times its radius r times π.

$C = \pi d$ or $C = 2\pi r$

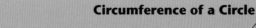

Examples

A. Find the circumference of the circle.

$C = \pi d$ circumference of a circle

$C = \pi \cdot 8.5$ Replace d with 8.5.

$C \approx 26.7$ ft   Use a calculator to simplify.

$9 \cdot \pi \approx 26.70353756$

The circumference is about 26.7 ft.

The circumference of a circle is the distance around the circle, while the area is the space inside the circle.

Area of a Circle

The area A of a circle is equal to π times the square of the radius r.

$A = \pi r^2$

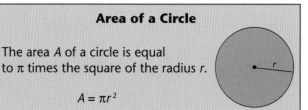

B. Find the area of the circle.

$A = \pi r^2$ area of a circle
$A = \pi \cdot 5^2$ Replace *r* with 5.
$A = \pi \cdot 25$ Evaluate 5^2.
$A \approx 78.5$ Use a calculator to simplify.

The area is about 78.5 cm².

Try THESE

Match the letter on the circle to the right with the parts listed below.

1. circumference
2. radius
3. diameter
4. chord
5. center

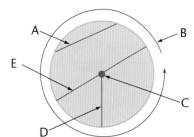

Exercises

Find the circumference and area of each circle. Round answers to the nearest tenth.

1.
2.
3.

4.
5.
6.

7. The radius is 6.8 in.

8. The diameter is 12.4 kilometers.

7.2 Circles

Find the area of each colored region. Round answers to the nearest tenth.

9.

10.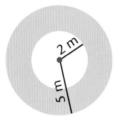

11.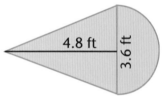

12.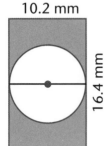

13.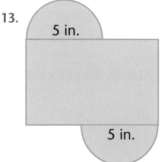

14.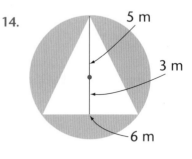

Problem SOLVING

15. What is the diameter of a circle if its circumference is 52.6 ft? Round your answer to the nearest tenth.

16. What is the radius of a circle if its area is 200 cm^2? Round your answer to the nearest tenth.

★17. A bicycle tire has a radius of 15.5 in. How far will the bicycle travel in 25 rotations of the tire? Round your answer to the nearest tenth.

Test PREP

18. What is the circumference of a circle with a radius of 8 ft?
    a. 50.24 ft^2
    b. 25.12 ft
    c. 25.12 ft^2
    d. 50.24 ft

19. What is the area of a circle with a diameter of 24 in.?
    a. 452.16 in.
    b. 452.16 in.^{2}
    c. 1,808.64 in.
    d. 1,808.64 in.^{2}

Cumulative Review

Estimate.

1. 4,329 + 8,837
2. 72.9 − 28.3
3. 5.48 × 6.1
4. 27)5,942
5. 6.2)1.79

Compute mentally.

6. 0.96)13.152
7. 23.7 • 100
8. 746 ÷ 1,000
9. 4 + -7
10. -6 + -17.5
11. -9 − 5
12. -16 − -1.2
13. 15 • -5
14. -21 • -0.1
15. 18 ÷ -6
16. -72 ÷ 8
17. -4.8 ÷ -12

Solve.

18. What number is 24% of 400?
19. What percent of 1,500 is 225?
20. 7 is 5% of what number?
21. 37.2% of 84 is what number?

Use the Pythagorean theorem to find the missing length for each right triangle.

22. $a = 6$ mm, $c = 10$ mm
23. $a = 9$ ft, $b = 40$ ft
24. $b = 8$ m, $c = 17$ m
25. $b = 9$ in., $c = 15$ in.
26. $a = 15$ in., $c = 39$ in.
27. $a = 16$ ft, $b = 30$ ft

Express each ratio to three decimal places.

28. sin J
29. cos J
30. tan J
31. sin L
32. cos L
33. tan L

Triangle JKL with right angle at K: JL = 7 in., LK = 2.695 in., JK = 6.461 in.

Evaluate each expression if $a = 4$, $b = -3$, and $c = \frac{1}{2}$.

34. $a + b$
35. $b - c$
36. ac
★37. $\frac{b}{c} + 1$

Solve.

38. Kathy telephones Fran four times a month. Each call costs the same amount. The total monthly cost is $7.40. How much does each call cost?

★39. Three men complete half of a job in 20 days. At this rate, how long will it take twelve men to complete the job?

40. On a map, 1 cm represents 125 km. The map distance between two cities is 2.5 cm. Find the actual distance.

41. A dealer pays Bonnie Gardner $2,000 for her car. The dealer sells the car for $2,400. What is the percent of increase?

42. Ben Jackson buys a microwave oven for $90.00 more than half of its original price. He pays $329.00 for the oven. What was its original price?

7.3 Three-Dimensional Figures

Objective: to identify and draw three-dimensional figures

Mr. Kocher is a land developer. He creates a three-dimensional drawing of a house as he plans a community.

A **three-dimensional figure**, or solid, is an object that does not lie in a plane. It has length, width, and height.
A **polyhedron** is a solid with flat surfaces that are polygons. Two examples of a polyhedron are prisms and pyramids.

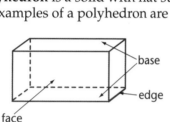

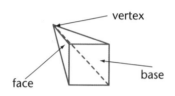

A **prism** has two parallel congruent faces called **bases**.

A **pyramid** has one base that is a polygon and faces that are triangles.

Prisms and pyramids are named by their bases. Two other types of solids are cylinders and cones.

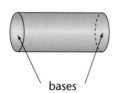

A **cylinder** has two bases that are parallel congruent circles.

A **cone** has exactly one circular base and one vertex.

Examples

A. Identify the solid. Name the shape of the faces. Name the number of faces, edges, and vertices.

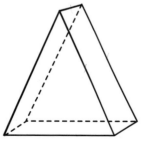

The solid is a triangular prism because it has two parallel congruent bases that are triangles.

There are three other faces that are rectangles.

It has a total of 5 faces, 9 edges, and 6 vertices.

B. Copy the figure to the right, and then draw the image that would be reflected across the line of symmetry.

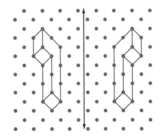

Try THESE

Identify each solid.

1.   2.   3.

Exercises

Identify each solid. Name the shape of the faces. Name the number of faces, edges, and vertices.

1.   2.   3.

Identify the two common solids that make up each structure.

4.    5.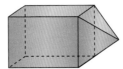

On isometric dot paper, copy each figure and the line of symmetry. Then draw the image that would be reflected across the line of symmetry.

6. 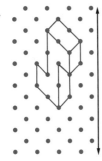 7. 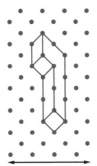 8.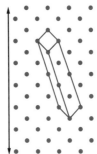

7.3 Three-Dimensional Figures

7.4 Volume of Prisms and Cylinders

Objective: to find the volume of prisms and cylinders

Volume is the amount of space that a solid occupies. Volume is measured in cubic units. One common unit is the cubic centimeter.

Volume: 1 cubic centimeter (cm³)

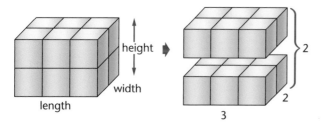

The volume (V) of a rectangular prism can be found by multiplying the length (l), width (w), and height (h).

What is the volume of the rectangular prism above?

The volume of the rectangular prism is 12 cubic centimeters.

$V = lwh$

$V = 3 \cdot 2 \cdot 2$

$= 12$

Examples

A. The volume of any prism can be found by multiplying the area of the base (B) and the height (h).

The formula is $V = Bh$.

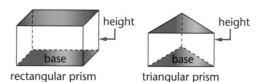

rectangular prism   triangular prism

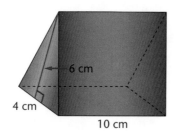

Find the volume of the prism shown at the left.

$V = Bh$

$V = \left(\frac{1}{2} \cdot 4 \cdot 6\right) \cdot 10$

$= 12 \cdot 10$

$= 120$

The base is triangular.

The area of a triangle is $\frac{1}{2} \cdot b \cdot h$.

The volume is 120 cm³.

B. The volume of a cylinder is also found by multiplying the area of the base (B) and the height (h). Since the area of a circle is πr^2, the formula is $V = \pi r^2 h$.

Find the volume of the cylinder shown.

$V = \pi r^2 h$

$V = 3.14 \cdot 6^2 \cdot 8$

$= 904.32$

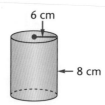

The volume is 904.32 cm³.

Try THESE

Find the volume of each prism and cylinder described below.
Use 3.14 for π.

1.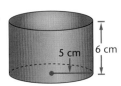
2.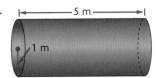
3.

Exercises

Find the volume of each figure described below.

1.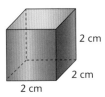
2.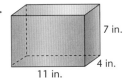
3.

4. rectangular prism: length = 37.1 ft, width = 25.9 ft, height = 64 ft

★5. hexagonal prism: base = 13.5 m², height = 7 m

6. cylinder: radius = 2.3 cm, height = 13.6 cm

7. cylinder: diameter = 15 in., height = 8 in.

Problem SOLVING

8. The volume of a square prism is 72 cm³. The edge of the base is 3 cm. What is the height?

9. Which has a greater effect on the volume of a cylinder, doubling the radius or doubling the height? Explain.

Mid-Chapter REVIEW

Find the area of each figure.

1.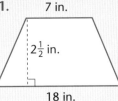
2.
3.

Find the area and circumference of each circle.

4.
5.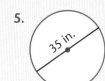

7.5 Volume of Pyramids and Cones

Objective: to find the volume of pyramids and cones

Dave Jaspers designs tents. He is designing a new model and deciding whether the top will be shaped like a pyramid or a cone. He decides to find the volume if each shape has the same height as the walls of the tent.

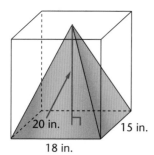

The volume of a pyramid is one-third the volume of a prism with the same base and height as the pyramid. The formula is $V = \frac{1}{3}lwh$.
Note: $lw = B$ (area of base)

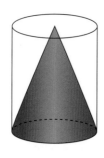

The volume of a cone is one-third the volume of a cylinder with the same radius and height as the cone. The formula is $V = \frac{1}{3}\pi r^2 h$.
Note: $\pi r^2 = B$ (area of base)

Examples

A. A pyramid has a length of 15 centimeters, a width of 12 centimeters, and a height of 10 centimeters. Find the volume.

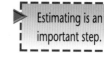

$V = \frac{1}{3}lwh$ Estimate: $15 \cdot 10 \cdot 10 \div 3 = 15 \cdot 100 \div 3$
$= 1{,}500 \div 3$
$15 \cdot 12 \cdot 10 \div 3 = 600$ $= 500$

$V = 600$ The volume is 600 cm³.

> Estimating is an important step.

B. Find the volume of the cone at the right.

$V = \frac{1}{3}\pi r^2 h$ Estimate: $\frac{1}{3}\pi \cdot 7^2 \cdot 18 \approx 1 \cdot 50 \cdot 20$
$\approx 1{,}000$

$V \approx \frac{1}{3} \cdot \frac{22}{7} \cdot \frac{7}{1} \cdot \frac{7}{1} \cdot \frac{18}{1}$

$V \approx 924$ The volume is about 924 ft³.

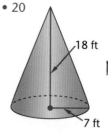

> Use $\frac{22}{7}$ for π when the radius is a multiple of 7.

Try THESE

Find the volume of each cone or pyramid described below. Use 3.14 for π.

1.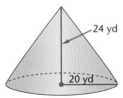
2.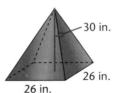
3.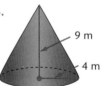

Exercises

Find the volume of each solid below.

1.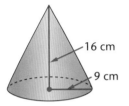
2.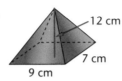
3.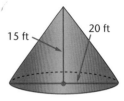

4. cone: radius = 6 yd, height = 8 yd

★5. cone: diameter = 29.2 m, height = 39.2 m

★6. pyramid: length = 8.2 in., width = 10 in., height = 15 in.

7.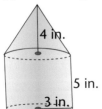
8.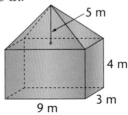

Problem SOLVING

9. A cone-shaped funnel has a radius of 4 inches and a height of $7\frac{1}{2}$ inches. Find the volume.

10. How are the formulas for the volume of a cone and the volume of a pyramid similar?

Mixed REVIEW

Compute.

11. 5 + -3
12. -12 + 5
13. 6 − 9
14. -2 − 1
15. 2.1 + -3.2
16. 4.1 + -1.4
17. 3.5 − 6.2
18. -8.1 − 5.4

7.6 Surface Area of Prisms and Pyramids

Objective: to find the surface area of prisms and pyramids

When you wrap a gift box, you completely cover all of the surfaces that make up the box. The **surface area** (S) of a polyhedron is the sum of the areas of all of the faces.

You can find the surface area of a prism by adding up all of the areas of the faces.

$$S = \text{top} + \text{bottom} + \text{front} + \text{back} + \text{side} + \text{side}$$
$$S = l \cdot w + l \cdot w + l \cdot h + l \cdot h + w \cdot h + w \cdot h$$
$$S = 2lw + 2lh + 2wh$$

To find the surface area of a square pyramid, you need to find the area of the base (a square) and the area of each face (four triangles), and then add.

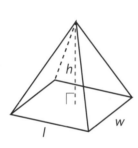
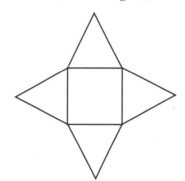

Area of faces: $A = \frac{1}{2} \cdot b \cdot h$ There are 4 triangles, so you need to multiply the area by 4.

Area of base: $A = l \cdot w$ Add the area of the faces and the area of the base.

$S = 4 \cdot \text{area of triangle} + \text{area of square}$

Examples

A. Find the surface area of the the rectangular prism.

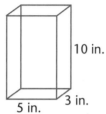

$S = 2lw + 2lh + 2wh$

$S = 2 \cdot 5 \cdot 3 + 2 \cdot 5 \cdot 10 + 2 \cdot 3 \cdot 10$

$S = 30 + 100 + 60$

$S = 190 \text{ in.}^2$

B. Find the surface area of the square pyramid.

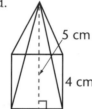

Area of faces = $4 \cdot \frac{1}{2} \cdot 4 \cdot 5 = 40$

Area of base = $4 \cdot 4 = 16$

Add. $40 + 16$

Surface area = 56 cm^2

Try THESE

1. Find the surface area of the rectangular prism.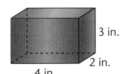

2. Find the surface area of the pyramid.

Exercises

Find the surface area of each solid.

1.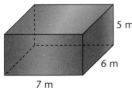

2.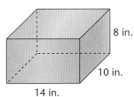

3.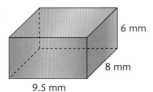

4.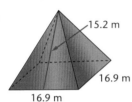

5.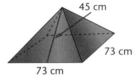

6.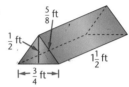

7. Jackie built a pyramid for a history project. The base is a square with each side measuring 12 inches. The height of each side is 15 inches. Find the surface area of the pyramid.

Constructed RESPONSE

8. Dave designed a tent with canvas sides and a canvas floor. Is 85 yd² of canvas enough to make the tent? (*Hint:* Use the Pythagorean theorem to find the height of the base.) Explain.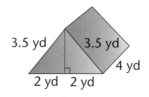

Mind BUILDER

Logical Thinking II

Volume, capacity, and mass are related in the metric system. One cm³ holds one mL of water, which weighs one gram.

volume = 1 cm³
capacity = 1 mL
mass = 1 g

1. How many milliliters of water have a mass of 65 grams?

2. What is the mass in grams of the water in a 3.4-m³ container?

3. How many liters of water have a mass of 825 grams?

4. What is the mass in kilograms of 8 liters of water?

7.6 Surface Area of Prisms and Pyramids

7.7 Surface Area of Cylinders and Cones

Objective: to find the surface area of cylinders and cones

The Jacksons have a swimming pool shaped like a cylinder. This summer, they want to put in a new plastic liner and cover.

Examples

A. To find the surface area of a cylinder, add the measures of the areas of its circular bases and its curved surface.

top	$\pi r^2 = \pi 5^2$	$= 3.14 \cdot 25$	$= 78.5$
bottom	$\pi r^2 = \pi 5^2$	$= 3.14 \cdot 25$	$= 78.5$
curved surface	$2\pi rh = 2\pi 5 \cdot 30$	$= 2 \cdot 3.14 \cdot 5 \cdot 30$	$= 942.0$

l (circumference $= 2\pi r$)

$S = 1{,}099.0$

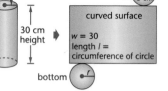

B. To find the surface area of a cone, add the area of its circular base and the area of its curved surface.

Find the surface area of the cone.
Use 3.14 for π.

> The area of the base is πr^2. The area of the curved surface is πrs.

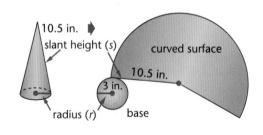

πr^2
base $3.14 \cdot 3 \cdot 3$ 28.26
curved surface $3.14 \cdot 3 \cdot 10.5$ $+\ 98.91$
πrs

$S \approx 127.17$ The surface area is about 127.17 in.^{2}.

Try THESE

Find the surface area of each solid.

1.
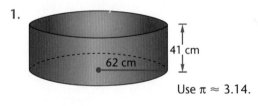
Use $\pi \approx 3.14$.

2.
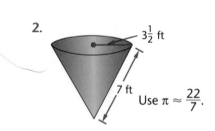
Use $\pi \approx \frac{22}{7}$.

194   7.7 Surface Area of Cylinders and Cones

Exercises

Find the surface area of each solid. Use 3.14 for π.

1.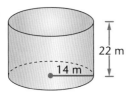
2.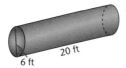
3.
★4.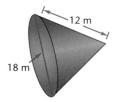

5. cylinder: diameter = 9 in., height = 15 in.
6. cone: diameter = 24 cm, height = 14 cm

Problem SOLVING

7. What is the surface area that frosting covers of a one-layer cake that has a diameter of 8 in. and a height of 2 in.? (*Remember:* You do not frost the bottom of the cake.)

8. How many square inches of metal are used to make a closed can measuring $2\frac{1}{2}$ in. wide and $4\frac{1}{2}$ in. tall?

★9. The area of the curved surface of a cone is π*rs*, where *s* is the slant height. A cylinder and a cone each have a radius of 3 in. and a height and slant height of 4 in. What is the ratio of the areas of their curved surfaces?

Test PREP

10. Which cylinder has the greatest surface area?
    a. radius = 4.5 m, height = 3 m
    b. radius = 3 m, height = 5 m
    c. radius = 3.5 m, height = 5 m
    d. radius = 5 m, height = 3 m

11. Which is the best estimate for the surface area of a cone with a radius of 8.5 in. and a slant height of 6 in.?
    a. 179 in.²
    b. 186 in.²
    c. 273 in.²
    d. 387 in.²

Mixed REVIEW

Rename each of the following as a decimal.

12. $\frac{1}{2}$
13. $\frac{1}{5}$
14. 38%
15. 145%
16. $2\frac{7}{10}$

Rename each of the following as a percent.

17. $0.3\overline{3}$
18. 0.375
19. $1\frac{1}{2}$
20. $\frac{5}{8}$
21. 1.3

Rename each of the following as a fraction or mixed number in simplest form.

22. 30%
23. 0.84
24. $0.6\overline{6}$
25. 320%
26. 2.125

7.8 Problem-Solving Strategy: Using a Diagram

Objective: to solve problems by using a diagram

A circular pond is in the center of Cloverdale Park. The diameter of the pond is 130 feet. The park is shaped like a trapezoid with a height of 260 feet and bases of 400 feet and 360 feet. Find the area of dry land in the park.

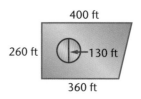

 Find the area of dry land. You know the dimensions of the circular pond and the park. The pond is inside the park.

 Make a diagram of the park and pond. Find the area of the pond. Find the total area of the park. Then subtract the area of the pond from the total area to find the area of dry land.

 area of pond

$A = \pi r^2$

$A = 3.14 \cdot 65 \cdot 65$

$A = 13,266.5$

The area of the pond is about 13,267 ft².

area of park

$A = \frac{1}{2}h(b_1 + b_2)$

$A = \frac{1}{2} \cdot 260(400 + 360)$

$A = 98,800$

The area of the park is 98,800 ft².

area of park − area of pond = area of dry land

98,800 − 13,267 = 85,533

The area of dry land is about 85,533 ft².

 The area of the pond plus the area of dry land should equal the total area.

    13,267
+ 85,533
    98,800 ✓

The answer is correct.

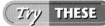

Solve by making a diagram. Use 3.14 for π.

1. The radius of a circular flower garden and its border is 2.8 meters. Without the border, the radius is 2.3 meters. Find the area of the border.

Solve

1. All but one corner of a rectangular field 180 m by 250 m is planted in soybeans. The corner is a right triangle 60 m by 60 m. Find the area planted in soybeans.

2. A steamroller is 3 feet wide, and its roller has a diameter of 2 feet. About how many square feet of pavement can it cover in one complete roll?

3. The can shown at the right is packed in a box 18 cm by 18 cm by 20 cm. Find the volume of space remaining in the box.

4. Does area, perimeter, or volume tell you the amount of wax in a candle?

5. If the radius of a candle is 6 m and the height is 9 m, what is the volume?

6. Cheryl melted wax that filled a container with a radius of 6 cm and a height of 23 cm. She wants to pour the wax into smaller molds that have a diameter of 7 cm and a height of 13 cm. How many of the smaller molds can Cheryl fill?

7. How would you use a ruler to find the volume of this textbook?

8. Find the volume of the barn shown at the right.

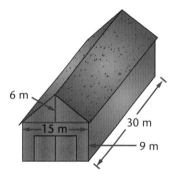

Chapter 7 Review

Language and Concepts

Choose a term or formula from the list at the right to complete each sentence.

1. A solid with surfaces (faces) that are polygons is a(n) _____.
2. A _____ is a polyhedron with two parallel congruent faces called bases.
3. The _____ of a polyhedron is the sum of the area of all the faces.
4. _____ is the amount of space that a solid contains.
5. The formula for the volume of a prism is $V =$ _____.
6. The formula for the volume of a cylinder is $V =$ _____.
7. $V = \frac{1}{3}lwh$ is the formula for the volume of a _____.
8. $V = \frac{1}{3}\pi r^2 h$ is the formula for the volume of a _____.

pyramid
volume
polyhedron
cone
Bh
prism
$\pi r^2 h$
surface area

Skills and Problem Solving

Find the area of each figure. Section 7.1

9.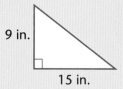
9 in.
15 in.

10.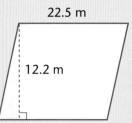
22.5 m
12.2 m

11.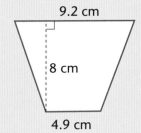
9.2 cm
8 cm
4.9 cm

Find the area and circumference of each circle. Round answers to the nearest tenth. Section 7.2

12.
8 ft

13.
$18\frac{1}{2}$ m

14.
22.4 in.

Find the volume of each solid. Use 3.14 for π. Sections 7.4–7.5

15.

16.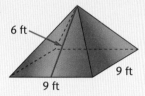

17.

Find the surface area of each solid. Use 3.14 for π. Sections 7.6–7.7

18.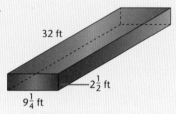

19.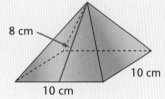

20.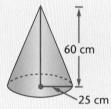

Solve. Use 3.14 for π. Sections 7.4–7.8

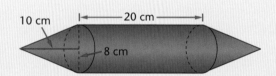

21. The machine part on the right is cylindrical with identical cones on each end. Find the volume.

22. A greenhouse is shaped like a triangular prism. Its base measures 64 square feet, and its height is 8 feet. What is the volume of the greenhouse?

23. A can of vegetables has a 3-in. radius and a 7-in. height. What is the approximate area of the label that covers the side of the can?

24. A cone-shaped funnel has a radius of 7 in. and a height of 10 in. What is the volume of the funnel to the nearest tenth?

25. A cone-shaped flour bin has a radius of 10 ft and a height of 18 ft. Find its volume to the nearest tenth.

Chapter 7 Test

Find the area of each figure.

1.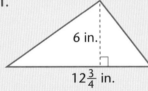
2.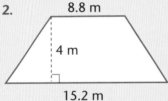
3.

Find the surface area of each solid. Use 3.14 for π.

4.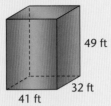
5.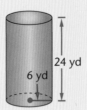
6.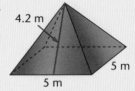

Find the volume of each solid. Use 3.14 for π.

7.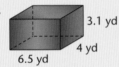
8.
9.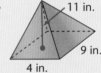

10. cylinder: base = 42 cm², height = 20 cm

11. pyramid: length = 24cm, width = 10 cm, height = 7 cm

12. cone: radius = 5 ft, height = 12 ft

13. prism: base = 93 cm², height = 26 cm

Solve. Use 3.14 for π.

14. A round block of cheese is 2.5 ft thick. The radius is 1.3 ft. Find its surface area.

15. A candle is 2 cm in radius and 10 cm tall. Find the amount of wax used in making it.

16. A circular part of a square field 600 m by 600 m is irrigated. The irrigated part has a radius of 300 m. Find the area of the field that is not irrigated.

200 Chapter 7 Test

Change of Pace

Spheres

A **sphere** is a solid with all points the same distance from a given point called the **center**. A line segment from any point on the sphere to the center is called the **radius** (r).

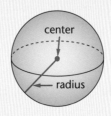

To find the surface area of a sphere, use the formula $S = 4\pi r^2$.

To find the volume of a sphere, use the formula $V = \dfrac{4\pi r^3}{3}$.

Find the surface area and volume of a sphere with a radius of 5 meters to the nearest whole number.

$S = 4\pi r^2$

$S \approx 4 \cdot 3.14 \cdot 5^2$

$S \approx 314$

The surface area is about 314 m².

$V = \dfrac{4\pi r^3}{3}$

$V \approx \dfrac{4 \cdot 3.14 \cdot 5^3}{3}$

$V \approx \dfrac{1{,}570}{3} \approx 523.\overline{3}$

The volume is about 523 m³.

Find the surface area and volume of each sphere described below. Round each answer to the nearest whole number.

1. 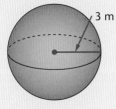 3 m

2. 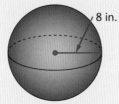 8 in.

3. 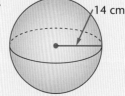 14 cm

4. radius = 60 km

5. radius = 710 km

6. radius = 940 ft

Solve. Use 3.14 for π.

7. The shape of Earth is almost a perfect sphere. The radius is about 6.4 megameters (million meters). Find the surface area to the nearest square megameter.

8. About 45 square megameters of land on Earth is used for farming. What percent of the surface area is used for farming? Round your answer to the nearest percent.

★9. Why does a sphere enclose more volume for a given surface area than any other shape?

Cumulative Test

1. What percent of 16 is 12?
   a. 25%
   b. 75%
   c. $133\frac{1}{3}\%$
   d. none of the above

2. Which of the following is the formula for the volume of a prism?
   a. $V = Bh$
   b. $V = \frac{1}{2}bh$
   c. $V = \pi r^2 h$
   d. $V = \frac{1}{3}Bh$

3. Which of the following parts of a circle is *not* a line segment?
   a. arc
   b. chord
   c. diameter
   d. radius

4. What is the cost of a $42.00 jacket at 20% off?
   a. $8.40
   b. $20.00
   c. $22.00
   d. $33.60

5. -8 + 7 = _____
   a. -15
   b. -1
   c. 15
   d. none of the above

6. Solve. $8y - 2 = 18$
   a. 2
   b. $2\frac{1}{2}$
   c. 3
   d. $4\frac{1}{4}$

7. What is the perimeter of the figure shown at the right?
   a. 28 cm
   b. about 32.56 cm
   c. about 45.12 cm
   d. none of the above

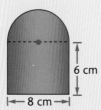

8. What is the area of the figure shown in problem 7?
   a. 48 cm²
   b. about 73.12 cm²
   c. 80 cm²
   d. about 98.24 cm²

9. A runner has won her first 4 races. She will run 12 races in all. What percent of the total races has she won?
   a. 4%
   b. 12%
   c. 25%
   d. $33\frac{1}{3}\%$

10. Using the following formula, which temperature is equivalent to -5°C?
    $F = \frac{9}{5}C + 32$
    a. -23°F
    b. 23°F
    c. 27°F
    d. 37°F

CHAPTER 8

Data and Statistics

Khadijeh Zarafshar
Illinois

8.1 Samples and Surveys

Objective: to understand samples and surveys

Advertisers study **statistical data** to find out what the "average person" likes. Statistical data are pieces of information that are gathered, organized, analyzed, and presented in numerical form.

When conducting a survey, you cannot ask everyone's opinion, so you take a **sample**. A sample is representative of a larger group, the population. In an **unbiased sample**, or random sample, each person or object has an equally likely chance of being selected. Three types of unbiased samples are listed below.

Type	Definition
Simple Random Sample	A sample where each person or object has an equally likely chance of being chosen.
Stratified Random Sample	A sample where the population is divided into smaller groups with certain characteristics. Then a simple random sample is selected from each group.
Systematic Random Sample	A sample where the people or objects are selected according to time intervals. (Every nth person or object is selected.)

In a **biased sample**, certain parts of the population are favored. It is not representative of the entire population. One type of biased sample is a **convenience sample**, which includes members of the population that are easily accessible. Another type is a **voluntary response sample**. This sample includes only those who want to participate in the sample.

Examples

A. Describe the sample.

To determine the most popular ice-cream flavor at an ice-cream parlor, every fifth person who entered the parlor was surveyed.

The sample is a systematic random sample because the population is the people in the ice-cream parlor. It is an unbiased sample.

B. A scarf store sells scarves in three colors: red, yellow, and black. Of the 20 people surveyed, they found that 6 preferred a red scarf, 4 preferred a yellow scarf, and 10 preferred a black scarf. What percent preferred the black scarf?

10 out of the 20 people preferred a black scarf.

$10 \div 20 = 0.50$ 50% of the people preferred a black scarf.

Try THESE

Tell whether each of these samples is a random sample.

1. A teacher is choosing a class representative and puts every student's name in a hat and chooses one.

2. You want to know the most popular movie, and you ask every fifth person who walks into the movie theater.

Exercises

Describe each sample.

1. To determine the most popular type of music listened to by students in school, a teacher asked the students in her first period class only.

2. To determine what store most people like in a mall, every ninth person who walked into the mall was surveyed.

3. To determine where the class should go on the eighth grade field trip, parents were asked to fill out and send in a questionnaire.

4. To determine what time school should start, two students from each grade were chosen at random and surveyed.

5. To evaluate the quality of the shoes a company makes, they tested every eighty-fifth pair of shoes.

6. To determine what sport people like the most, the spectators at a football game were surveyed.

Problem SOLVING

The student government is planning a school ice-cream social. One hundred members of the student body were asked to state what flavor of ice cream they will eat, and the results are in the table to the right.

Flavor	Number
Chocolate	42
Vanilla	25
Strawberry	12
Cookie Dough	21

7. What percent of the students prefer chocolate ice cream?

8. What percent of the students prefer cookie dough ice cream?

9. If 350 students attend the ice-cream social, about how many students will eat vanilla ice cream?

Mixed REVIEW

Solve each equation.

10. $\frac{1}{3} \cdot \frac{2}{5} = x$

11. $\frac{5}{8} \cdot t = \frac{25}{32}$

12. $b \cdot 3\frac{1}{5} = 8$

13. $8\frac{1}{3} \cdot r = 2\frac{1}{2}$

14. $\frac{4}{9} - \frac{1}{3} = a$

15. $c - 1\frac{1}{4} = \frac{1}{2}$

16. $2\frac{3}{5} + s = \frac{4}{5}$

17. $p - 3 = 2\frac{3}{5}$

8.1 Samples and Surveys

8.2 Circle Graphs

Objective: to construct and interpret circle graphs

A **circle graph** is used to compare parts of a whole. The whole amount is shown as a circle. Each part is shown as a percent of the whole. Use the steps listed below to make a circle graph.

Step 1 Label the graph.

Step 2 Use a compass to draw a circle.

Step 3 Multiply 360° by each percent to find the angle for each part. Round to the nearest degree.

Step 4 Use a protractor to draw the angles by placing the center of the protractor at the center of the circle.

Step 5 Label each part of the circle.

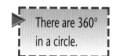
There are 360° in a circle.

Example

Make a circle graph for the data in the table at the right.

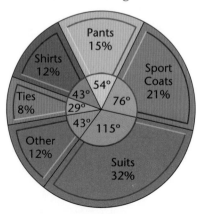

Sale of Men's Clothing	
Suits	32%
Sport Coats	21%
Pants	15%
Shirts	12%
Ties	8%
Other	12%

32% of 360° = 0.32 • 360 ≈ 115°

21% of 360° = 0.21 • 360 ≈ 76°

15% of 360° = 0.15 • 360 ≈ 54°

12% of 360° = 0.12 • 360 ≈ 43°

8% of 360° = 0.08 • 360 ≈ 29°

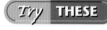

Use the circle graph in the Example above to answer each question.

1. What percent of the sales is shirts?

2. Which item provided approximately one-third of the total sales?

3. The sales for suits are how many times greater than the sales for ties?

4. The sales for pants are about one-half of the sales of what other item?

Exercises

Make a circle graph for the data in each problem.

1.
Davis Family Budget	
Category	Percentage
Housing	30%
Taxes	25%
Food	20%
Transportation	11%
Miscellaneous	9%
Savings	5%

2.
Chemical Composition of the Human Body	
Element	Percentage
Oxygen	65%
Carbon	18%
Hydrogen	10%
Nitrogen	3%
Other	4%

3.
Investing Money	
Type of Investment	Percentage
Stocks	41%
Bonds	12%
Savings Accounts	18%
Mutual Funds	29%

4.
Movie Attendance	
Day	Percentage
Monday	4%
Tuesday	6%
Wednesday	9%
Thursday	12%
Friday	25%
Saturday	29%
Sunday	15%

5.
Types of Vehicles Owned	
Type	Percentage
Sedan	45%
Wagon	12%
SUV	27%
Minivan	16%

6.
Money to Build a House	
Source	Percentage
Savings and Loans	38.8%
Commercial Banks	23.5%
Life Insurance	18.8%
Mutual Savings Banks	16.3%
Pension Funds	2.6%

Problem SOLVING

Use the table to the right displaying the medals won by the U.S. in the 2006 Winter Olympics in Torino, Italy, to solve each problem.

7. What percentage of the medals won were gold? silver? bronze?

8. Draw a circle graph to represent the data.

9. Use the circle graph to describe the medals won by the U.S.

Medals Won by the U.S. in the 2006 Winter Olympic Games	
Medal	Number
Gold	9
Silver	9
Bronze	7

8.3 Central Tendency

Objective: to find the mean, median, and mode of a set of data

Meg asked her classmates the following question, "About how many hours did you talk on the telephone last week?" She tallied their answers in a frequency table and then made a **bar graph**.

Hours	Tally	Frequency
0	III	3
1	III	3
2	IIII	4
3	IHI	5
4	III	3
5	I	1
6	I	1

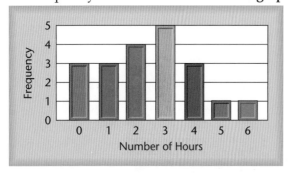

Which number of hours occurs the most? Which occurs the least?

What type of data is easiest to find in the bar graph?

Meg wants to find a number to describe the set of data. A **measure of central tendency** is a number that represents the center of a set of data. Mode, mean, and median are measures of central tendency.

Examples

A. The **mode** is the number that occurs most often.
What is the mode of the number of hours on the telephone?

The mode is 3 hours because it occurs most often.

B. The **mean**, or average, is the sum of the data divided by the number of pieces of data.

What is the mean of the number of hours on the telephone?

hours • frequency
$0 \cdot 3 = 0$
$1 \cdot 3 = 3$
$2 \cdot 4 = 8$
$3 \cdot 5 = 15$
$4 \cdot 3 = 12$
$5 \cdot 1 = 5$
$6 \cdot 1 = 6$
   20   49 ← total number of hours

sum of data ↙   ↘ number of pieces of data

$49 \div 20 = 2.45$
The mean is 2.45 hours.

C. The **median** is the middle number when the data are arranged in order. When there are two middle numbers, the median is their mean.

What is the median of the number of hours on the telephone?

0, 0, 0, 1, 1, 1, 2, 2, 2, 2, 3, 3, 3, 3, 3, 4, 4, 4, 5, 6

Since there are 20 pieces of data, the median is the mean of the tenth and eleventh pieces of data.

The median is $\frac{2+3}{2}$ or 2.5 hours.

Try THESE

Find the mean, median, and mode for each set of data.
Round to the nearest tenth.

1. 14, 13, 10, 9, 13, 6, 5
2. 41, 45, 53, 49, 57, 38, 50
3. 3.4, 1.8, 2.6, 1.8, 2.3, 3.1

Exercises

Copy and complete each frequency table. Then find the mean, median, and mode for the data, if possible.

★1.

Height	Tally	Frequency
5'2"	IIII	4
5'3"	IIII IIII	
5'4"	IIII IIII IIII	
5'5"	IIII III	
5'6"	III	

2.

Favorite Sport	Tally	Frequency
soccer	IIII I	6
baseball	IIII IIII	
basketball	IIII IIII IIII	
football	IIII IIII	
other	IIII IIII I	

Problem SOLVING

3. Garcia's mean score on his three French tests this month is 77. His first two scores are 75 and 79. He forgets his third score. What is his third score?

4. Seven employees earn $10,000, three earn $12,000, and two earn $40,000. Find the mean, median, and mode of the salaries.

5. Make a frequency table and a bar graph of the following math test scores. Use intervals of 5. Then find the mean, median, and mode of the data set.

   98, 88, 81, 77, 74, 69, 94, 85, 75, 72, 85, 79, 72, 65, 88, 82, 78, 74, 70, 62

Constructed RESPONSE

6. Which is affected more by a large or a small number, the mean or the median? Explain.

7. Write a situation that has a set of data in which the median better represents the data than the mean does.

Test PREP

8. In the following list of data, which number is the median?

   21, 28, 31, 35, 39, 43, 51, 60

   a. 35 b. 37 c. 38.5 d. 40

9. The following is the set of scores Ivy earned on her science tests. What is the mean of the test scores if her teacher does not count the lowest test score?

   86, 71, 94, 88, 82, 95

   a. 86 b. 71 c. 85 d. 89

8.3 Central Tendency

8.4 Stem-and-Leaf Plots

Objective: to organize data with stem-and-leaf plots

Jay Oney checks the inventory at a record store. Each day he marks the number of records sold.

The amounts sold each day in a 3-week period were 28, 30, 17, 31, 22, 33, 45, 44, 35, 28, 33, 19, 49, 39, 47, 34, 32, 36, 21, 28, and 42.

Numerical data can be organized in a **stem-and-leaf plot**. The greatest place value of the data is used for the stem. The next greatest place value is used for the leaves.

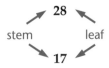

stem   leaf

The stem-and-leaf plot below shows the data that Mr. Oney collected.

The data range is from 17 to 49. So, the stems range from 1 to 4.

Stem	Leaf
1	7 9
2	1 2 8 8 8
3	0 1 2 3 3 4 5 6 9
4	2 4 5 7 9

1|7 = 17

Record the units digit, or leaf, next to the correct stem. Order the leaves from least to greatest.

Use the stem-and-leaf plot above to answer the following questions.

1. What is the least number of records sold in 1 day?

2. What is the greatest number of records sold in 1 day?

3. Are there any clusters or gaps in the data?

4. What does 3|0 represent?

5. What is the mode of the data?

6. What is the median of the data?

7. Each horizontal group of numbers represents an interval of 10. For example, the first line 1|7 9 represents 10–19. In which interval do most days of sales fall?

Exercises

State the stems that you would use to plot each set of data.

1. 17, 32, 41, 34, 42, 35
2. 9, 12, 24, 51, 33, 14, 18, 26
3. 294, 495, 272, 153, 240, 427
4. 7.5, 5.4, 8.6, 6.3, 7.1, 5.9, 8.2

Find the median and mode of the data in each stem-and-leaf plot.

5.
Stem	Leaf
4	2 7
5	0 3 4 7 9
6	1 5 6 8
7	1 7 8

4|2 = 42

6.
Stem	Leaf
8	1 2 2
9	2 4 6 9
10	0 3 7 8 9

8|1 = 81

7.
Stem	Leaf
2	0 1 2 5 7 7
3	1 3 4
4	5 6 7 8 9

2|0 = 20

The data at the right are the ages of people who attended a play. Use this data to solve each problem.

25, 33, 15, 7, 18, 24, 15, 25, 17, 26, 42, 35, 16, 26, 11, 22, 9, 13, 23, 16, 19, 7, 12, 18, 17, 6, 14, 38, 24, 29

8. Construct a stem-and-leaf plot of the data.

9. What was the age of the youngest person attending the play?

10. What was the age of the oldest person attending the play?

11. With what age group was the play most popular?

Problem SOLVING

12. Use the data at the right to construct a stem-and-leaf plot. In what height interval are most student council members?

Heights in Inches of Student Council Members

55, 62, 66, 64, 70, 59, 63, 65, 58, 72, 68, 60, 63, 68, 72, 59, 62, 67, 60, 61

★13. Jill weighs $\frac{3}{4}$ as much as Jack. If Jill holds a 20-pound weight, she balances Jack. How much does Jack weigh?

★14. Find the approximate perimeter and area of the figure below.

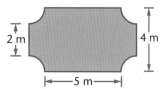

8.5 Measures of Variation

Objective: to find the range and quartiles for each set of data

Sandy Dennison has a choice of moving to Los Angeles or to Dallas for her job. She likes warm weather. So, she compares the temperatures in both cities.

She compares the data by finding the **measures of variation**. A measure of variation describes the spread of values in a set of data.

The **range** is the difference between the greatest and least numbers. What are the ranges in temperature for Los Angeles and Dallas?

The **interquartile** is the middle half of the data. An ordered set of data can be divided into four equal parts called **quartiles**. The **interquartile range** is the difference between the lower and upper quartiles of the data. An **outlier** is a value that is 1.5 times the value of the interquartile range.

Mean Daily Temperatures (°F)		
Month	Los Angeles	Dallas
January	57	44
February	59	49
March	60	56
April	62	66
May	65	74
June	69	82
July	74	86
August	75	86
September	73	79
October	69	68
November	63	56
December	58	48

Example

Find the interquartile range of the mean daily temperatures in Los Angeles.

1. Find the median of the data.

   57 58 59 60 62 63 | 65 69 69 73 74 75
                                     median

   The median is 64. Why? $\frac{63 + 65}{2} = 64$

2. Find the median of each half.

   57 58 59 | 60 62 63 | 65 69 69 | 73 74 75
          lower   upper
        quartile   quartile

   The lower quartile is $\frac{59 + 60}{2}$ or 59.5. The upper quartile is $\frac{69 + 73}{2}$ or 71.

The interquartile range is 71 − 59.5, or 11.5. The middle half of the mean daily temperatures in Los Angeles vary 11.5°F.

 **THESE**

Use the information presented above to answer each question.

1. What is the median of the temperatures in Dallas?
2. What is the lower quartile of the temperatures in Dallas?
3. What is the upper quartile of the temperatures in Dallas?
4. What is the interquartile range of the temperatures in Dallas?

Exercises

Find the range, lower quartile, upper quartile, interquartile range, and any outliers for each set of data.

1. 4, 16, 12, 16, 18, 18, 20, 22
2. 5, 27, 30, 31, 35, 38, 39, 40, 42
3. 12, 21, 23, 25, 33, 37, 38, 39, 42, 42
4. 2, 10, 14, 21, 23, 23, 25, 32, 38, 64
5. 82, 85, 86, 88, 90, 92, 94, 103, 107, 119

Problem SOLVING

Refer to the table on p. 212 to solve problems 6–8.

6. Which city has a greater range of temperatures?
7. Compare the medians of the mean daily temperatures of the two cities.

Constructed RESPONSE

8. What do the interquartile range and range reveal about the Los Angeles and Dallas temperatures that the mean does not? Explain.

Mind BUILDER

Mean Variation

The mean variation of a set of data is the average amount each number differs from the mean.

Use exercises 1–4 below to find the mean variation of the following temperatures (°C): 28°, 23°, 27°, 25°, 22°, and 24°.

1. Find the mean of the temperatures to the nearest tenth.
2. Find the difference between each number and the mean.
3. Add the differences.
4. Divide the sum by the number of addends to find the mean variation.
5. What kinds of changes would greatly affect the mean variation?

Problem Solving

Pentominoes

Pentominoes are arrangements of five congruent squares that meet at their edges. Four pentominoes are shown below.

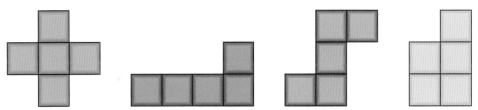

Use graph paper to make the twelve different pentominoes (including the ones shown above). Arrange the twelve pentominoes to make each of the following rectangles.

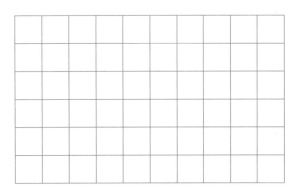

6 by 10

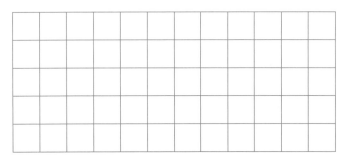

5 by 12

15 by 4

Extension

Now make a square piece from four squares. Using the twelve pentominoes and the square piece, make an 8-by-8 square.

Cumulative Review

Estimate.

1. $3.21 + 6.84$
2. $483 - 78$
3. 72×39
4. $73\overline{)296}$
5. $\$1.89 \times 4$
6. $\frac{3}{4} + \frac{5}{6}$
7. $6\frac{1}{5} - 5\frac{3}{10}$
8. $7\frac{1}{2} \div 1\frac{3}{4}$
9. $2\frac{1}{6} \cdot \frac{1}{2}$

Solve each equation.

10. $x - 4 = -7$
11. $-5y = -35$
12. $\frac{x}{6} = -4$
13. $x + 5 = 2$

Evaluate each expression if $x = 4$, $y = 6$, and $z = \frac{1}{3}$.

14. $xy + z$
15. $3z + \sqrt{x}$
16. $y^2 - 6z$
17. $\frac{4x - y}{z}$

Compute. Write each answer in simplest form.

18. $\frac{1}{3} + \frac{1}{9}$
19. $6 - \frac{7}{9}$
20. $\frac{4}{9} \cdot \frac{3}{8}$
21. $\frac{4}{5} \div \frac{8}{9}$
22. $\frac{7}{8} - \frac{1}{3}$
23. $3\frac{1}{3} + 2\frac{4}{5}$
24. $5\frac{1}{4} \div \frac{7}{8}$
25. $\frac{5}{7} \cdot 1\frac{2}{5}$
26. $1\frac{2}{3} \cdot 2\frac{1}{4}$
27. $9\frac{1}{4} - 3\frac{2}{3}$
28. $8\frac{6}{7} + \frac{2}{3}$
29. $4\frac{1}{4} \div 2\frac{1}{8}$

Use the figure at the right to name each of the following.

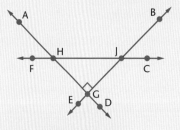

30. two perpendicular line segments
31. a right triangle
32. two right angles
33. two obtuse angles

Use the stem-and-leaf plot to answer each question.

Stem	Leaf
1	7 9
2	1 2 2 3 5 7 9 9
3	0 0 0 2 4 4 5 8
4	0 5

1|7 = 17

34. What is the greatest number in the data?
35. What is the range of the data?
36. What is the median of the data?

Solve.

37. Jordan is a baseball fan. Five times last season he bought a $3.50 seat, four times a $4.50 seat, and two times a $5.50 seat. What was the mean price he paid?

38. Today, Julie's Bakery sold 18 more cakes than what were sold yesterday. In both days combined, 54 cakes were sold. How many cakes were sold yesterday?

Cumulative Review 215

8.6 Box-and-Whisker Plots

Objective: to display and interpret data in a box-and-whisker plot

The heights of the boys on the Madison Middle School basketball team are 63 inches, 64 inches, 65 inches, 65 inches, 66 inches, 66 inches, 67 inches, 68 inches, 69 inches, and 70 inches. How can this information be displayed?

You can draw a **box-and-whisker plot** to show the median, quartiles, and interquartile range of the data.

- Draw a horizontal line.
- Plot the median, the quartiles, and the extremes (lowest and highest points).
- Draw a *box* around the middle half of the data. The interquartile range is 3 inches.
- Mark the median by drawing a line through its point.
- Draw the *whiskers* by connecting the lower quartile to the lower extreme and the upper quartile to the upper extreme.

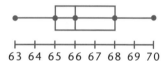

Try THESE

Use the box-and-whisker plot at the right to answer each question.

1. What is the median?
2. What is the lower quartile?
3. What is the upper quartile?
4. What is the interquartile range?
5. What is the lower extreme?
6. What is the upper extreme?
7. What is the range of the data?

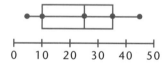

Exercises

Answer the following questions about the box-and-whisker plot to the right.

1. What is the greatest data point?
2. Between what two data points is the middle half of the data?
3. What is the range of the data?
4. What part of the data is less than 25?

Make a box-and-whisker plot for each set of data.

5. 32, 38, 27, 29, 21, 25, 41, 47, 23, 51, 22
6. 165, 173, 154, 143, 161, 152, 185, 150, 160, 175
7. 14.8, 12.3, 16.1, 12.9, 8.5, 6.9, 7.5

Problem SOLVING

Use the box-and-whisker plot below to solve each problem.

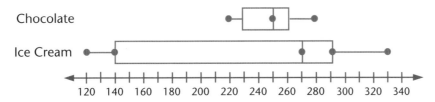

Calories in One-Serving Snack Foods

8. Compare the range of both sets of data.
9. What percent of ice cream has more than 270 calories?
10. What percent of chocolate has more than 230 calories?

Mid-Chapter REVIEW

Solve using the data in the chart to the right.

1. Find the mean, median, and mode.
2. Find the lower and upper quartiles and the interquartile range.
3. Draw a box-and-whisker plot of the data.

Scores on a History Quiz
24 12 19 24 20 14 23 15
16 16 24 12 17 22 13 17
17 16 17 19 20 23 15 17

8.6 Box-and-Whisker Plots   217

8.7 Choosing an Appropriate Display

Objective: to choose an appropriate display for a set of data

When you have a set of data, you have many choices in how you can display it. Some of the ways to display data are shown below.

Display	Definition
Bar Graph	A graph using bars to show comparisons between different values.
Box-and-whisker plot	A way to show the median, quartiles, and interquartile range of the data.
Circle Graph	A graph using a circle to compare parts of a whole.
Histogram	A graph which shows the frequency of data arranged by intervals.
Line Graph	A graph in which line segments are used to compare changes in data over time.
Line Plot	A picture of information on a number line. An **x** is often used to indicate the occurrence of a number.
Pictograph	A graph that uses pictures to display data.
Stem-and-leaf plot	A way to organize data that displays all numbers in the data set by place value.
Table	A way to list data individually or by groups.

As you choose how to display your data, ask yourself what do you want your graph to display and what kind of data do you have to display.

Example

Choose an appropriate type of data display. The table below shows a city's population from 1985–2005.

Year	Population
1985	78,550
1990	81,520
1995	76,040
2000	80,680
2005	83,500

A line graph would be an appropriate display for this data because it is used to compare quantities that are constantly changing.

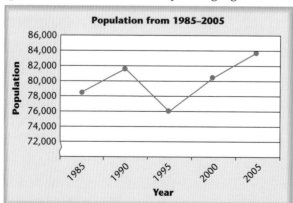

Try THESE

Complete the following.

1. Compare and contrast line plots and line graphs.
2. Compare and contrast circle graphs and bar graphs.

Exercises

Choose an appropriate data display for each situation, and explain your reasoning.

1. a survey of your friends' favorite types of music
2. the hourly temperatures throughout the day
3. the percent of income spent on monthly expenses
4. the heights of various buildings in a city
5. the sales of a brand of pasta over the course of 10 years
6. the average scores on a history test categorized by intervals

Problem SOLVING

The following are two ways to display the test scores received on Mr. Alexander's math test. Use these displays to solve each problem.

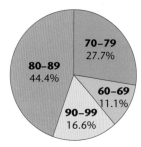

7. Can you tell from the circle graph how many students earned an 85?
8. Can you tell from the stem-and-leaf plot how many students earned between 80 and 89?
9. Compare and contrast the stem-and-leaf plot to the circle graph with respect to the math test scores.

8.8 Misleading Graphs

Objective: to recognize when graphs and statistics are misleading

An advertising agency develops two versions of a graph for a George's Pizza ad. Do both graphs show the same information? What makes them appear so different? Which graph do you think George's Pizza would prefer? Why?

Sometimes using a different scale helps to make a graph easier to read. Other times, scales are chosen to make the data appear to support a particular point of view.

Examples

A. Advertisements may use graphs that have no scales or only one scale. This graph seems to show statistical results but actually gives no information.

B. A graph may appear to show a positive association between two factors, even though the factors are probably not related.

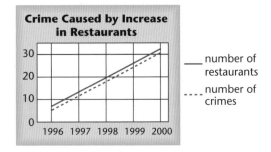

Use the graph in Example A to answer each question.

1. What fraction of repairs do "Ours" appear to have compared to "Theirs"?

2. If the vertical scale is labeled 9, 10, 11, and 12, what fraction of repairs do "Ours" have compared to "Theirs"?

Exercises

Use the bar graph at the right to answer each question.

1. How many more tickets did the ninth graders sell than the seventh graders?

2. By looking only at the bars, it appears that the ninth graders sold how many times more tickets than the seventh graders?

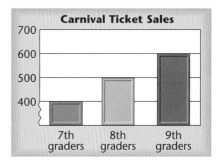

Use the graph at the left to answer each question.

3. What happened to the sales from last month to this month?

4. How do the volumes of the two figures compare?

5. How is this graph misleading?

Problem SOLVING

6. Explain how the statistical graph at the right is misleading if gas costs are actually one-fifth of nuclear costs.

7. **Research**
   Look in newspapers or magazines to find a graph that you think is misleading. Explain why you think it is misleading.

★8. One calorie of energy is required to heat 1 g of water 1°C. How many calories are required to raise 10 g of water from 3°C to 8°C?

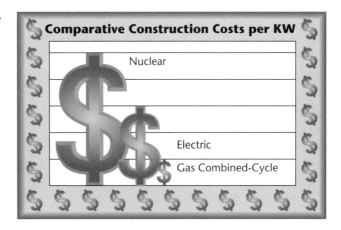

Mixed Review

Solve each equation.

9. $4r - 2 = 30$

10. $0.2b + 0.6 = 1.4$

11. $2.1 + 3m = 17.4$

12. $13 = 2 + \dfrac{c}{5}$

13. $\dfrac{p}{3} + 8 = 12$

14. $\dfrac{s}{4} - 9 = 2.3$

8.8 Misleading Graphs

8.9 Scatter Plots

Objective: to construct and interpret scatter plots

Silvia wants to know if the amount of education a person completes affects the amount of salary she or he receives.

She plots the data from the chart at the right onto a **scatter plot**, as shown below.

A scatter plot is a graph of two variables. The horizontal scale represents one variable, and the vertical scale represents the other.

Education and Income, U.S.A.		
Years of School Completed	Average Annual Income	
	Women	Men
Less than 8 years	$10,153	$14,485
8 years	11,183	18,541
9–11 years	12,267	20,003
12 years	15,947	24,701
College —		
1–3 years	18,516	28,025
4 or more years	24,482	36,665

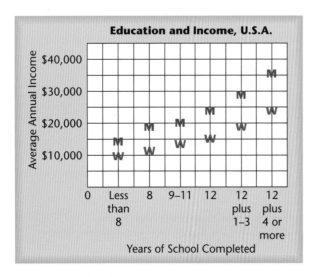

Sometimes you can see a **correlation**, or trend, between the two variables on a scatter plot. This scatter plot shows a **positive correlation** between the variables of schooling and income.

How would a scatter plot with a **negative correlation** appear?

If the points are scattered without an obvious trend, there is no correlation.

Try THESE

What type of correlation is shown by each of the following scatter plots?

1. Hourly Wage vs. Age

2. Gas Mileage of Car vs. Age of Car

3. Test Score vs. Study Time

Exercises

State whether a scatter plot for the following examples would show *positive correlation*, *negative correlation*, or *no correlation*.

1. outside temperature, heating bill
2. height, weight
3. test scores, height of student
4. miles per gallon, weight of car
5. sales by a salesperson, years of experience of salesperson
6. amount of money earned, amount of money spent

Problem SOLVING

7. Draw a scatter plot for the data given below. How are the two groups of data related?

Hours Worked	12	23	18	46	17	18	34	15	21	10	40	1	21
Money Earned ($)	51	82	65	199	157	68	186	75	189	42	180	5	170

8. Using the data in problem 7, find the mean number of hours worked and the mean amount of money earned. Then compare.

9. Find a statistical graph in a newspaper. Write a problem that can be solved using the graph.

★10. Two calendar dates are 180 days apart. If the first date is on a Tuesday, on what day of the week is the second date?

Mind BUILDER

Percentile

Test Scores

98 ← 99th percentile
88
87 ← 75th percentile
82
80 ← 50th percentile
76
75 ← 25th percentile
67

On various nationwide tests, a **percentile** is used to describe the score. For example, the eighty-second percentile is the point at or below which 82% of the scores fall. Suppose your test score is 87. Your score is at the seventy-fifth percentile. This means that 75% of the scores are at or below your score.

Solve. Use the list above.

1. Suppose your score is 80. At what percentile is your score?
2. Suppose your score is 98. At what percentile is your score?
3. Add 95, 69, 84, and 95 to the list. What score is at the seventy-fifth percentile?
4. Suppose your score is 75. At what percentile is your score?

8.10 Problem-Solving Strategy: Looking for a Pattern

Objective: to find patterns in data sets and graphs

Mr. Cummings thinks that the number of hours students watch television affects their grade point average. So, he took a survey to find the number of hours his students watched TV each week.

Boys' Results

Hours of TV	10	20	0	30	35	25	10	28
Grade Point Average	2.8	2.25	3.6	1.5	1.5	1.9	1.8	1.9

Girls' Results

Hours of TV	25	0	8	18	15	3	7	15	18
Grade Point Average	2.25	3.1	3.3	3.0	2.7	3.4	2.9	2.25	3.3

Is Mr. Cummings' assumption correct? Does there appear to be a correlation in the data?

You know the number of hours of television watched each week and the grade point average of each student. You need to determine if there is a relationship between the two factors.

Since there are two variables—hours of TV watched and grade point average—you can show the data on a scatter plot. Represent each boy with a **B** and each girl with a **G**.

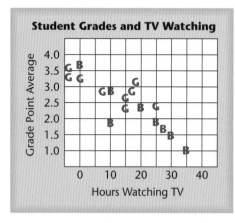

The results seem to show a negative correlation. It appears that the more TV a student watches, the lower the grade point average.

The conclusion that watching more hours of TV causes lower grade point averages appears to be true.

Caution! Statistics cannot *prove* that one factor *causes* another to happen. Statistics can only show that a relationship *may* exist. What other reasons might explain the results shown on the scatter plot?

Try THESE

Use the histogram at the right to solve the following problem.

1. The histogram shows the heights of all eighth grade boys at Jackson Middle School. The curve formed by connecting the centers of the bars is similar to a *normal* or *bell curve*. Name at least two other sets of data that have this pattern.

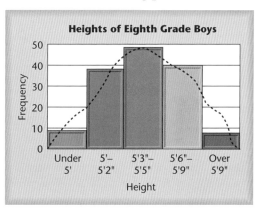

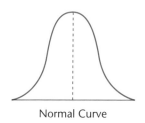

Solve

Use the double-line graph to answer each question.

1. What was the life expectancy for men born in 1920? for women born in 1920?

2. In what year was the difference between life expectancy for men and women the least?

3. Between which years did life expectancy show the most increase?

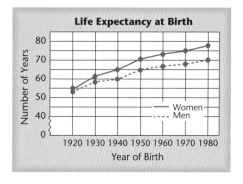

★4. Describe the trend shown in life expectancy for both men and women.

★5. Describe the trend shown in the difference between men's and women's life expectancies.

Use the data shown below to solve each problem.

School Enrollment Since 1960

	1960	1965	1970	1975	1980	1985	1990	1995	2000
Ashton	502	490	510	550	542	538	560	510	518
Sells	396	398	425	480	512	540	565	570	581

6. Make a double-line graph for the data.

7. How do the changes in enrollment differ between the two schools?

8. What might be the reasons for the differences?

★9. Can you predict any future trends from this graph?

8.10 Problem-Solving Strategy: Looking for a Pattern

Chapter 8 Review

Language and Concepts

Choose the correct term to complete each sentence.

1. The middle point of an ordered set of data is the _____.
2. Data can be divided into four equal parts called _____.
3. The _____ is the number that occurs most often.
4. In a(n) _____ sample, a person has an equally likely chance of being selected.
5. The average of a set of data is the _____.
6. A(n) _____ sample includes people who are easily accessible.

range
sample
unbiased
median
mode
mean
convenience
voluntary
quartiles

Skills and Problem Solving

Describe each sample. Section 8.1

7. To determine what pizza topping people at a local pizzeria like best, every fifth person who entered the pizzeria was surveyed.

8. To determine what sport middle school students like, the boys on the basketball team were surveyed.

Make a circle graph for the set of data below. Section 8.2

9.

Water Use in the U.S.	
Department	Percentage
Agriculture	36%
Public Water	8%
Utilities	33%
Industry	23%

Find the mean, median, and mode for each set of data. Section 8.3

10. 156, 175, 212, 167, 628, 156

11. 2.6, 3.1, 6.8, 4.9, 5.7, 3.4, 4.3

12. For problem 11 above, which best describes the data, the mean or the median?

Use the data at the right to solve each problem. Sections 8.4–8.5

13. Make a stem-and-leaf plot of the data.
14. What is the greatest movie attendance?
15. What is the median of the data?
16. What are the upper and lower quartiles?
17. What is the interquartile range?

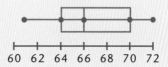

Movie Attendance
25 32 17 28 30 78 19
35 30 21 25 34 39   7

Use the box-and-whisker plot to answer each question. Section 8.6

18. What is the range of the data?
19. What is the median of the data?
20. What fraction of the data is less than 64?

Choose an appropriate data display for each situation, and explain your reasoning. Section 8.7

21. the populations of cities in the Northeast arranged by intervals
22. the average monthly temperature for a year

Use the graphs below to answer each question. Section 8.8

23. Which graph could be used to indicate a greater difference in the number of books read?
24. What is the difference between the two graphs?

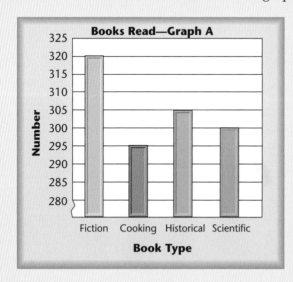
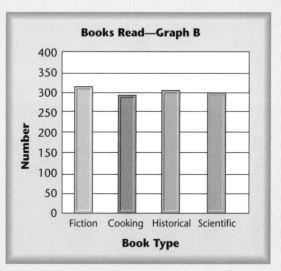

State whether a scatter plot for the following examples would show *positive correlation, negative correlation,* or *no correlation.* Section 8.9

25. practice time, number of points scored in a basketball game
26. time, temperature of hot chocolate

Chapter 8 Review   227

Chapter 8 Test

Describe each sample.

1. To determine the best destination for a vacation, a travel show asked people to call in and complete a survey.

2. To determine how often people travel by air, the members of an airlines frequent flyer club were surveyed.

Make a circle graph for the data set below.

3.

Elements in the Earth's Crust	
Element	Percentage
Oxygen	47%
Silicon	28%
Aluminum	8%
Iron	5%
Other	12%

Use the data below of the number of hours of television watched weekly by a randomly selected group of people to solve problems 4–7.

Weekly Television Watching Survey

Hours	20	19	19	34	21	46	32	49	38	25	40	19	41	25	25	28	32	33	22	36

4. What are the mean, median, and mode for the number of hours of television watched weekly?

5. What is the range of the hours?

6. What is the interquartile range?

7. Make a box-and-whisker plot of the hours.

8. Would a scatter plot of age and hours of television watched weekly have a *positive correlation*, *negative correlation*, or *no correlation*?

Solve.

9. Miguel scored 85, 78, and 90 on three of his four science tests. His average is 83. What was his fourth score? Show all work, and explain.

10. Describe at least two misleading factors often found in graphs.

11. What are the lower quartile, median, upper quartile, and interquartile range of the data given at the right?

```
35  37  28  21  45  36  18  24
42  35  27  32  49  51  47  19
22  33  31  20  38  42  26  21
```

228   Chapter 8 Test

Change of Pace

Cumulative Frequency Histograms

Mrs. Shin made the histogram of history test results shown at the right. She wanted to know how many students scored 80 or less, so she made the **cumulative frequency histogram** below.

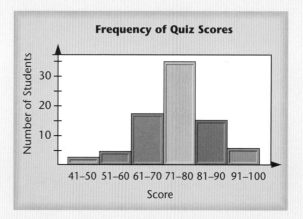

The number of scores in each interval is added to the number of scores in the interval before it. Each bar is built by adding to the top of the bar before it. Therefore, each bar shows the number of scores that fall below the score at the right vertical boundary of that bar.

The **cumulative relative frequency** (%) is the cumulative frequency divided by the total number of test takers.

$$\frac{\text{cumulative frequency}}{\text{total number of test takers}} = \%$$

If 60 students scored 80 or less, the cumulative relative frequency is:

$$\frac{60 \div 20}{80 \div 20} = \frac{3}{4} \quad \blacktriangleright \quad \frac{3 \cdot 25}{4 \cdot 25} = \frac{75}{100} \quad \blacktriangleright \quad 75\%$$

Therefore, 75% of all students who took the test scored 80 or less.

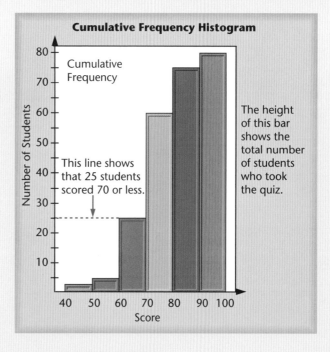

Use the data shown below of mathematics test scores to solve each problem.

73	88	68	91	79	88	73	90	75	72
65	85	81	81	78	96	80	59	86	87
81	78	87	83	87	80	77	75	86	84

1. Construct a cumulative frequency histogram of the mathematics test scores.
2. How many scores are there?
3. How many scores are 70 or below?
★ 4. Find the cumulative relative frequency of scores 70 or below.
★ 5. What is the cumulative relative frequency of scores 80 or below?

Cumulative Test

1. Which two decimals are equivalent to 0.1?
   a. 0.01 and 0.001
   b. 1.0 and 10
   c. 0.10 and 0.100
   d. none of the above

2. Solve.
   $$\frac{w}{9.1} = 27.3$$
   a. 3
   b. 18.2
   c. 248.43
   d. none of the above

3. Which group of numbers is arranged in order from least to greatest?
   a. 0.18, 0.081, 0.81, 1.08
   b. 0.081, 0.18, 0.81, 1.08
   c. 0.18, 0.81, 0.081, 1.08
   d. 1.08, 0.81, 0.18, 0.081

4. Which is the sum of the measures of the angles in a triangle?
   a. 90°
   b. 180°
   c. 360°
   d. none of the above

5. Which temperature is the mean for the data below?
   25° 26° 28°
   26° 24° 21°
   a. 24°
   b. 25°
   c. 26°
   d. none of the above

6. Triangle ABC can be classified as which of the following?
   a. equilateral, acute
   b. isosceles, acute
   c. scalene, obtuse
   d. none of the above

7. Mary bought a gallon of milk at $1.29 and two packages of lunch meat at $1.69 each. What was the total cost?
   a. $2.98
   b. $4.27
   c. $4.77
   d. none of the above

8. Which angle has a degree measure of about 75?

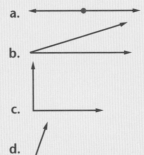

9. The graph shows the growth of a plant. How tall do you think the plant will be on the fifth day?
   a. 5 cm
   b. 6 cm
   c. 7 cm
   d. 7.5 cm

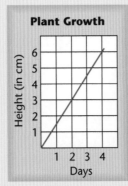

10. Bonita sold 14 candles for $2.50 each. She paid $21.00 for materials to make the candles.

    Which of the following can be the first step to finding the amount of profit Bonita made?
    a. 14 • $2.50
    b. $2.50 + $21.00
    c. $21.00 − 14
    d. $21.00 − $2.50

CHAPTER 9

Probability

Alexander Hormozi
Calvert Day School

9.1 Probability

Objective: to find the probability of an event

Rosalita and Fernando are playing a game. Two number cubes are rolled. Rosalita wins if the product is an even number. Fernando wins if the product is an odd number. Is this game fair or unfair?

The 36 outcomes in the **sample space** are listed in the multiplication table.

×	1	2	3	4	5	6
1	1	2	3	4	5	6
2	2	4	6	8	10	12
3	3	6	9	12	15	18
4	4	8	12	16	20	24
5	5	10	15	20	25	30
6	6	12	18	24	30	36

> The outcome 5, 3 results in a product of 15. The loops on the table show this product.

There are 9 odd products, shown in color, and 27 even products.

> **Probability** is the ratio of the number of ways an event can occur to the total number of possible outcomes.

An **event** is a specific outcome such as "an odd product."

The probability of an odd product is found as follows.

$$P(\text{odd product}) = \frac{9}{36} \quad \leftarrow \text{number of successful outcomes} \\ \quad\quad\quad\quad\quad\quad\quad\quad \leftarrow \text{number of possible outcomes}$$

> Read $P(\text{odd product})$ as the "probability of an odd product."

The probability of an odd product is $\frac{9}{36}$ or $\frac{1}{4}$.

Examples

A. What is $P(\text{product} \geq 30)$? Refer to the multiplication table above.

Three products 30, 30, and 36 are greater than or equal to 30. So,

$$P(\text{product} \geq 30) = \frac{3}{36} \text{ or } \frac{1}{12}.$$

The probability of rolling an even product is $\frac{3}{4}$. The events of rolling an even product and rolling an odd product are **complementary events**. The sum of the probabilities of complementary events is 1.

> You can find the probability of two events that cannot happen at the same time. These two events are called **mutually exclusive**.
>
> $P(\text{event A or event B}) = P(\text{event A}) + P(\text{event B})$

B. Find the probability of rolling a 6 or a number less than 4 with a single cube.

$P(6 \text{ or less than } 4) = P(6) + P(\text{less than } 4)$

$$= \frac{1}{6} + \frac{3}{6}$$

$$= \frac{4}{6}$$

The probability of rolling a 6 or a number less than 4 is $\frac{4}{6}$ or $\frac{2}{3}$.

> To find the probability of two events that are not mutually exclusive, add the probabilities, and then subtract the event that happens at the same time.
>
> P(event A or event B) = P(event A) + P(event B) − P(event A and B)

C. Find the probability of rolling a 3 or an odd number.

$$P(3 \text{ or odd number}) = P(3) + P(\text{odd number}) - P(3 \text{ and odd})$$
$$= \frac{1}{6} + \frac{3}{6} - \frac{1}{6}$$
$$= \frac{3}{6} \text{ or } \frac{1}{2}$$

The probability of rolling a 3 or an odd number is $\frac{1}{2}$.

Try THESE

List the sample space for each of the following.

1. tossing a coin
2. rolling a number cube
3. spinning the spinner

Exercises

A number cube is rolled. Find each probability.

1. $P(\text{number} > 2)$
2. $P(\text{prime number or 5})$
3. $P(\text{number} \geq 3)$
4. $P(\text{factor of 6 or 5})$
5. $P(\text{even number or odd number})$
6. $P(\text{multiple of 3 or 6})$
7. In Exercises 1–6, which have mutually exclusive events?
8. Are the events of rolling a number greater than 3 and rolling a number less than 3 complementary?

A card is drawn from a deck of 26 cards. Each card is printed with a different letter of the alphabet. Find the probability of drawing each of the following.

9. **M** or **N** or **O**
10. a vowel or a consonant
★11. any letter in *Louisiana* or any letter in *Alabama*

Problem SOLVING

12. Two number cubes are rolled. What is the probability that a sum of 12 is rolled? What is the probability that the sum is not 12?
13. What is the probability that a month picked at random has 31 days?
14. A bag has 6 green marbles, 4 red marbles, and 5 yellow marbles. The probability of picking a purple marble is $\frac{1}{6}$. How many purple marbles are in the bag?

9.2 Counting Outcomes

Objective: to use tree diagrams and the fundamental counting principle

Casey is ordering a milk shake from the Malt Shop. The flavors to choose from are chocolate, vanilla, strawberry, or coffee. They come in small or large. She can get it with or without a cherry. How many ways can Casey order a milk shake?

You can make a **tree diagram** to find the number of combinations or outcomes.

Flavor   **Size**   **Cherry**

- Chocolate
  - Small
    - cherry
    - no cherry
  - Large
    - cherry
    - no cherry
- Vanilla
  - Small
    - cherry
    - no cherry
  - Large
    - cherry
    - no cherry
- Strawberry
  - Small
    - cherry
    - no cherry
  - Large
    - cherry
    - no cherry
- Coffee
  - Small
    - cherry
    - no cherry
  - Large
    - cherry
    - no cherry

There are 16 ways Casey can order a milk shake.

You can also use the **fundamental counting principle** to find the number of milk shakes. The fundamental counting principle is used to find the total number of possible outcomes of an event.

If event A can occur in m ways and for each of these ways, an event B can occur in n ways, then events A and B can occur in $m \cdot n$ ways.

Number of flavors • Number of sizes • Cherry or not = Total number of milk shakes

     4   •   2   •   2   =   16

Example

When wrapping a gift, you have 8 rolls of wrapping paper, 6 ribbons, and 5 cards to choose from. How many ways can you wrap a gift?

       wrapping paper • ribbon • card

           8   •   6   •   5   fundamental counting principle

There are 240 ways to wrap a gift.

Try THESE

Draw a tree diagram to find the number of possible outcomes.

1. A number cube is rolled, and a quarter is tossed.
2. You can get a hamburger or cheeseburger with French fries or onion rings and a small, medium, or large soft drink.
3. You can have a cup of mint, orange, or cinnamon tea with or without sugar.

Exercises

Use the fundamental counting principle to find the number of possible outcomes.

1. A penny is tossed five times.
2. There are four choices for ten questions on a multiple-choice test.
3. There are roses, tulips, and carnations in red, yellow, white, pink, purple, or orange.
4. A pair of jeans comes in blue or black. They come in small, medium, or large.
5. A girl's school uniform has two skirts, three sweaters, and four shirts.

Problem SOLVING

6. You are shopping for a new car. A dealer offers the model you want in a 2- or 4-door sedan. It comes in blue, black, grey, red, white, or silver. You can get an automatic or manual transmission. How many cars do you have to choose from?

Constructed RESPONSE

7. A dress store sells an A-line dress in 15 different colors and high-heel shoes in 15 different colors. Sara said there are 30 different ways to wear a dress with a pair of shoes.
   a. What error did she make?
   b. What is the correct answer?
   c. Explain your reasoning.

Test PREP

8. A bike comes in five different colors with the option of two kinds of racing stripes and three kinds of handlebars. How many bike choices are there?

   a. 10   b. 8   c. 30   d. 20

9.3 Permutations

Objective: to find the number of permutations of objects

Luis, Tom, and Joan play trumpet for the concert band. In how many ways can they be seated in the three chairs?

A list of all 6 possible seating arrangements is given.

Another way to find the number of seating arrangements is to use multiplication. Imagine that Luis, Tom, and Joan sit in these chairs.

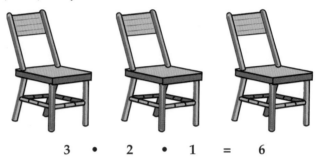

> Any one of the 3 can sit in the first chair. Either of the remaining 2 can sit in the second chair. Only 1 remains to sit in the third chair.

3 • 2 • 1 = 6

Three students can be seated in 3 • 2 • 1, or 6, different orders.

You can write 3 • 2 • 1 as 3! Read 3! as "three factorial." Numbers, like 3! and 9!, are **factorials**. The factorial of a number is the product of all the whole numbers, except 0, that are less than or equal to that number.

An arrangement in which order is important is called a **permutation**.

There are 20 members of the music honor society. They need to choose a president and a vice president. They have 20 members to choose from for president and 19 to choose from for vice president. There are 20 • 19, or 380, ways to choose a president and vice president. This can be written as the permutation $_{20}P_2$.

A permutation can be written as $_nP_r$, where n represents the number of objects chosen r at a time.

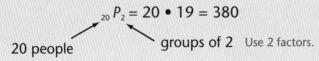

Examples

A. Simplify. $_{12}P_3$

$_{12}P_3 = 12 \cdot 11 \cdot 10$ product of three factors starting with 12

$= 1{,}320$ Simplify.

B. Four people are running in a race. How many ways can they finish the race?

There are 4 people who can finish first, 3 who can finish second, and so on. This is the same as 4!

4! = 4 • 3 • 2 • 1 = 24

There are 24 ways four people can finish the race.

Try THESE

Simplify each expression.

1. 6!
2. 9!
3. 3! + 2!
4. 6! − 3!

Exercises

Simplify each expression.

1. $_9P_3$
2. $_{28}P_2$
3. $_{50}P_3$
4. $_{250}P_2$

5. Five friends are having their picture taken. How many ways can the photographer arrange them in a row?

6. How many ways can six books be arranged in a row on a shelf?

7. How many ways can you arrange the letters in the word *equation*?

Problem SOLVING

8. Eight students are in the final of the Math Olympiad. How many ways can they finish in first, second, and third place?

Constructed RESPONSE

9. A student simplified 5! and found the answer to be 15. What did the student do wrong? What is the correct answer? Explain.

10. Lynn and Leonard completed the problem to the right. Who completed the problem correctly? Explain.

Lynn
$_6P_3 = 6 • 5 • 4$
$= 120$

Leonard
$_6P_3 = 6 • 5 • 4 • 3 • 2 • 1$
$= 720$

Mixed REVIEW

Solve each equation.

11. $x − -7 = 9$
12. $25 = -5x$
13. $\dfrac{x}{-9} = -6$
14. $15 = x + -9$
15. $x − -8 = -7$
16. $-4x = -64$

9.4 Combinations

Objective: to find the number of combinations of objects

The concert band has three trombone players—Ann, Bill, and Chang. How many ways can two soloists be chosen?

Ann, Bill	Ann, Chang	Bill, Chang
~~Bill, Ann~~	~~Chang, Ann~~	~~Chang, Bill~~

You can cross out the duplicates. Two soloists can be chosen in 3 ways. The order in which you choose two people does not matter. This is an example of a **combination**. A combination is a group of items in which order is not important.

> A combination can be written as $_nC_r$, where n represents the number of objects chosen r at a time.
>
> $$_nC_r = \frac{_nP_r}{r!} \qquad _3C_2 = \frac{_3P_2}{2!} = \frac{3 \cdot 2}{2 \cdot 1} = 3$$

Example

The state symphony has thirty-five members. Two of the members are going to speak to the concert band. How many ways can two members of the state symphony be chosen?

This is a combination because order is not important.

$$_{35}C_2 = \frac{_{35}P_2}{2!} = \frac{35 \cdot 34}{2 \cdot 1} = \frac{1190}{2} = 595$$

There are 595 ways two members can be chosen from thirty-five.

State whether each example is a *permutation* or a *combination*.

1. 6 students in a row in math class

2. 4 out of 6 students are chosen for the math contest team

3. 3 books in a row on a shelf

4. a committee of 5 chosen from a group of 9

Exercises

Simplify each expression.

1. $_9C_2$
2. $_5C_1$
3. $_{10}C_9$
4. $_{12}C_1$
5. $_7C_5$
6. $_8C_3$
7. $_{15}C_6 + {}_3C_2$
8. $_{11}C_5 + {}_6C_2$

State whether each example is a *permutation* or a *combination*.

9. selecting a batting order in a baseball game
10. a pizzeria allows you to choose 2 toppings from 8 for your pizza
11. choosing 5 starting players from a team of 12
12. stacking 5 books on a table

Problem SOLVING

13. A class of 30 students needs to elect 3 student representatives. How many ways can 3 students be selected from 30?

14. You are at a frozen yogurt shop. How many ways can you choose 3 toppings from 21?

15. Your school can send 2 students to the national conference of middle school students. How many ways can 2 students be selected from 150?

Constructed RESPONSE

16. You are making a salad at a salad bar with 15 toppings.

    a. How many ways can you choose 7 toppings from the 15?

    b. You want to make a layered salad and order of ingredients is important. How many ways can you make a layered salad with 7 toppings from the 15?

    c. Why is there a difference between the number of ways to make the non-layered and layered salad?

Test PREP

17. Which is not a combination?

    a. choosing 3 toppings for your pizza

    b. lining 3 students up in a row

    c. 3 student representatives are selected from a class of 25

    d. choosing 2 desserts from a dessert tray

18. How many ways can you choose 3 dishes from 8 on a Chinese food menu?

    a. 336
    b. 56
    c. 24
    d. 144

9.4 Combinations

9.5 Pascal's Triangle

Objective: to identify patterns in Pascal's triangle

Pascal's triangle is named after the French mathematician Blaise Pascal (1623–1662).

Although this triangle of numbers was known for centuries before Pascal was born, he was the first person to apply it to probability.

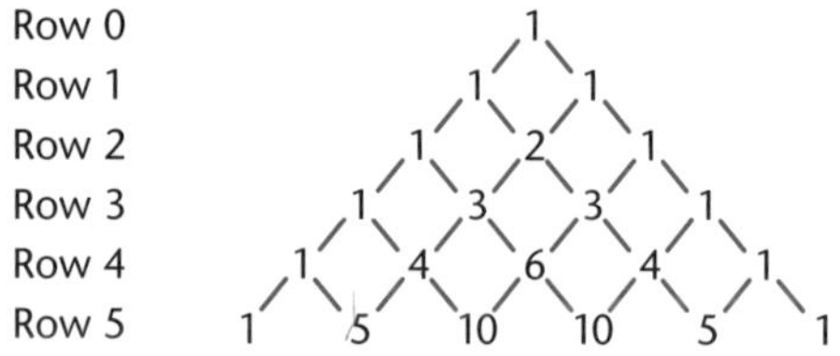

```
Row 0                    1
Row 1                  1   1
Row 2                1   2   1
Row 3              1   3   3   1
Row 4            1   4   6   4   1
Row 5          1   5   10  10   5   1
```

Exploration Exercise

1. Make a copy of Pascal's triangle, row 0 through row 5. Be sure to leave enough room for at least five more rows.

2. Determine how the numbers in row 3 can be used to find the numbers in row 4.

3. Extend Pascal's triangle to row 10.

4. Do you think there is a limit to the number of rows?

5. Add the numbers in each row of Pascal's triangle. Describe the pattern of sums. Write the sums as powers. Write the sum of the nth row as a power.

Row	0	1	2	3	4	5	6	7	8	9	10
Sum	1	2	?	?	?	?	?	?	?	?	?

6. Triangular numbers were presented in the Mind Builder on p. 3. Find the pattern of triangular numbers in Pascal's triangle.

> The number of combinations can be found in Pascal's triangle using the row number and column number.
>
> row number ⟶ nC_r ⟵ column number

You can use Pascal's triangle to find a number of combinations.

Example

There are four runners in a qualifying race. Any number of runners can qualify for the finals if their time is under 1 minute. But, if all their times are more than 1 minute, no one will qualify. Suppose you call the runners **A**, **B**, **C**, and **D**. List the combinations for each number of qualifiers in the following chart.

Number of Qualifiers	zero	one	two	three	four
$_nC_r$	$_4C_0$	$_4C_1$	$_4C_2$	$_4C_3$	$_4C_4$
Number of Combinations	1	4	6	4	1

The number of combinations looks like the fourth row of Pascal's triangle. The row number is the number of runners. The column number is the number of qualifiers.

Try THESE

Use Pascal's triangle to find each combination.

1. $_2C_1$
2. $_4C_0$
3. $_5C_3$

Exercises

Use Pascal's triangle to find each combination.

1. $_5C_1$
2. $_3C_2$
3. $_7C_4$
4. $_2C_0$
5. $_5C_5$
6. $_6C_3$

Problem SOLVING

Use a row of Pascal's triangle to help you solve each problem.

7. When 3 coins are tossed, there are $1 + 3 + 3 + 1$, or 8, possible outcomes. Compute the number of possible outcomes for 8 coins.

8. A coin is tossed 6 times. What is the probability of tossing exactly two heads?

9. A quiz has 10 true-false questions. The correct answers, in order, are **T, T, F, F, F, F, T, T, F,** and **T**. What is the probability you will earn 100% on the quiz, just by guessing?

Mid-Chapter REVIEW

A number cube is rolled. Find each probability.

1. $P(4)$
2. $P(\text{an even number})$
3. $P(3 \text{ or a factor of } 4)$

Simplify.

4. $_8P_3$
5. $_{25}C_2$

9.5 Pascal's Triangle

9.6 Theoretical and Experimental Probability

Objective: to find theoretical and experimental probability

Exploration Exercise

1. Make a spinner with four sections as shown to the right.
2. Spin the spinner 50 times.
3. Record the results in a chart like the one below.

Color	Occurrence
Red	
Blue	
Yellow	
Green	

4. Compute the ratio $\dfrac{\text{number of times color was spun}}{\text{total number of spins}}$ for each color.
5. Based on the spinner, what is the probability of the spinner landing on red, blue, yellow, or green?
6. Is this the same as your results? Explain why or why not.

By spinning the spinner, you performed an experiment. When you perform an experiment, you find the **experimental probability**. Experimental probabilities can vary each time the experiment is performed.

To find the theoretical probability, you do not need to perform an experiment. The **theoretical probability** is based on characteristics or known facts. The spinner above is divided into four equal parts. The probability of the spinner landing on one of the parts is $\frac{1}{4}$.

Theoretical Probability = $P(\text{event}) = \dfrac{\text{number of favorable outcomes}}{\text{total number of possible outcomes}}$

The experimental probability gets closer to the theoretical probability the more times the experiment is repeated.

Example

When you roll a number cube 30 times, how many times would you expect to roll a 6?

$$P(6) = \frac{1}{6} \cdot 30 = 5$$

You would expect to roll a 6 five times.

Try THESE

Complete the following.

1. If you were to toss a coin 100 times, what is the theoretical probability the coin will land on heads?

2. Toss a coin 100 times. What is the experimental probability of heads?

3. Compare your theoretical and experimental probability from problems 1–2.

Exercises

The table to the right shows Megan Farrington's season statistics for baseball. Use the table to answer each question.

Home Runs	5
Triples	3
Doubles	3
Singles	13
Walks	8
Outs	50
Plate Appearances	82

1. What is P(home run)?

2. What is P(single)?

3. How many walks would you expect Megan to earn in her next 50 plate appearances?

4. How many hits would you expect Megan to earn in her next 25 plate appearances?

Problem SOLVING

5. If you were to draw a card, then replace it, and repeat this process 50 times, how many Xs would you expect to draw?

6. If you were to spin the spinner 200 times, how many times would you expect to land on a 7?

7. If you were to draw a marble from the bag 150 times, and replace it after each draw, how many yellow marbles would you expect to draw?

8. The probability of a rainy day in Houston is about 0.285. Estimate the number of rainy days per year in Houston.

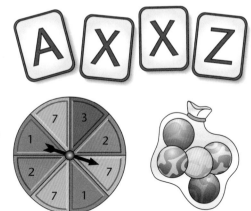

Constructed RESPONSE

9. A bag contains 25 red gumballs and 32 blue gumballs. You select a gumball from the bag at random.

   a. What is the theoretical probability of picking a red gumball?

   b. What is the theoretical probability of picking a blue gumball?

   c. The table to the right shows the results of picking a gumball, recording the color, and then returning the gumball to the bag. Find the experimental probability of each color gumball based on the table.

Color	Occurrence
Red	15
Blue	22

9.6 Theoretical and Experimental Probability

Problem Solving

Name That Cube

The number 370 is special because it is equal to the sum of the cubes of its digits.

$$3^3 + 7^3 + 0^3 = 27 + 343 + 0 = 370$$

Find two other numbers less than 500 that are equal to the sum of the cubes of their digits. (There are five including 370.)

Extension

Study the pattern shown at the right. Write the cubes of 4 and 5 as the sum of odd numbers.

$1^3 = 1$

$2^3 = 3 + 5$

$3^3 = 7 + 9 + 11$

Cumulative Review

Write in standard form.

1. $(6 \cdot 10^3) + (8 \cdot 10^2)$
2. $\sqrt{100}$
3. $\sqrt{1}$

Compute mentally. Write only your answers.

4. $4.8 - 0.9$
5. $438 + 909$
6. $27 \cdot 40$
7. $2{,}600 \div 13$

Evaluate.

8. $12 - b^2$, if $b = 0$
9. $9.71 + \sqrt{a}$, if $a = 100$

Solve each equation.

10. $b + 7 = 19$
11. $71 = y - 46$
12. $5c = 80$
13. $\frac{w}{3} = 12$

14. State what to work first and what to work second to solve $3a + 25 = 67$. Then solve the equation.

Compute. Write each answer in simplest form.

15. $\frac{8}{15} + \frac{4}{15}$
16. $\frac{5}{6} + \frac{2}{9}$
17. $4\frac{3}{4} + 3\frac{1}{3}$
18. $\frac{11}{12} - \frac{5}{12}$

19. $\frac{5}{8} - \frac{1}{2}$
20. $12\frac{4}{5} - 2\frac{1}{10}$
21. $\frac{2}{3} \cdot \frac{4}{5}$
22. $2\frac{1}{4} \cdot \frac{2}{3}$

23. $\frac{4}{21} \cdot 1\frac{3}{4}$
24. $\frac{2}{7} \div 3$
25. $\frac{9}{16} \div 2\frac{1}{10}$
26. $1\frac{3}{5} \div 4\frac{1}{5}$

Use the box-and-whisker plot to answer each question.

27. What is the range of the data?
28. What is the median?
29. Between which two numbers do half of the data lie?
30. In which quartile would 14 lie?

Solve.

31. Suppose your test scores are 90, 80, and 85. What is your average score? Which method of computation did you use to solve?

32. Triangle ABC is equilateral. Find the value of x.

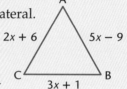

33. Shelly has a math quiz every fourth school day and a spelling quiz every third school day. The last date she had both quizzes was Tuesday, December 1. Find the next day and date she will have both quizzes.

9.7 Independent and Dependent Events

Objective: to find the probability of independent and dependent events

Carla has a bag of marbles. She picks one marble out of the bag, replaces it, and then picks another marble. If she does not replace the first marble before picking the second marble, do you think the probability will be different?

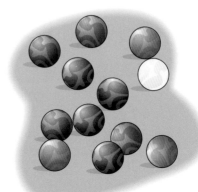

Picking two marbles out of a bag one at a time is an example of a **compound event**. A compound event consists of two or more simple events. Picking a marble from a bag, replacing it, and then picking another marble are examples of **independent events**. The outcome of the first event does not affect the outcome of the second event.

> The probability of two independent events can be found by multiplying the probability of the first event by the probability of the second event.
>
> P(event A and event B) = P(event A) • P(event B)

When the outcome of the first event affects the outcome of the second event, the events are called **dependent events**.

> The probability of two dependent events can be found by multiplying the probability of the first event by the probability of the second event after the first event has occurred.
>
> P(event A and event B) = P(event A) • P(event B after event A)

Examples

A bag has 7 blue marbles, 5 red marbles, and 3 yellow marbles.

A. Two marbles are picked one at a time from the bag. When a marble is picked, it is replaced. What is the probability of picking a red marble and then a blue marble?

P(red, blue) = P(red) • P(blue)

$= \dfrac{5}{15} \cdot \dfrac{7}{15}$ Since the first marble was replaced, picking the second marble is not affected.

$= \dfrac{7}{45}$ Multiply, and write the answer in simplest form.

246   9.7 Independent and Dependent Events

B. Two marbles are picked one at a time from the bag. When a marble is picked, it is not replaced. What is the probability of picking two yellow marbles?

$P(\text{yellow, yellow}) = P(\text{yellow}) \cdot P(\text{yellow})$

$= \dfrac{3}{15} \cdot \dfrac{2}{14}$ ← Number of yellow marbles after one marble is removed
← Number of marbles in the bag after one marble is removed

$= \dfrac{1}{35}$ Multiply, and write the answer in simplest form.

Try THESE

State whether each set of events is *independent* or *dependent*.

1. Two spinners are spun.
2. Two socks are removed from a drawer one at a time and are not replaced.
3. Pick a card. Replace it. Then pick another card.
4. Spin a spinner, and pick a card.

Exercises

A number cube is rolled, and a spinner is spun. Find each probability.

1. $P(5 \text{ and blue})$
2. $P(\text{odd and } \mathbf{C})$
3. $P(\text{a multiple of 2 and } \mathbf{A})$
4. $P(3 \text{ and } \mathbf{D})$
5. $P(\text{less than 4 and red})$
6. $P(\text{a factor of 3 and } \mathbf{C})$

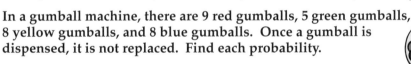

In a gumball machine, there are 9 red gumballs, 5 green gumballs, 8 yellow gumballs, and 8 blue gumballs. Once a gumball is dispensed, it is not replaced. Find each probability.

7. $P(\text{green, yellow})$
8. $P(\text{blue, blue})$
9. $P(\text{red, blue})$
10. $P(\text{blue, yellow, green})$
11. $P(\text{red, red, red})$
12. $P(\text{two gumballs that are neither red nor green})$

Problem SOLVING

13. What is the difference between independent and dependent events? Explain.

★14. There are 7 socks in a drawer in 3 different colors. The probability of picking 2 blue socks at random and not replacing them is $\frac{1}{7}$. How many blue socks are in the drawer?

9.8 Odds

Objective: to find the odds of events

Lisa plays center for the Rosemont Middle School basketball team. During the season, she has made 35 out of 50 free throws. Based on her free throw results during the season, Lisa can find the odds that she will make her next free throw.

The **odds in favor** of an event are found using this ratio:

$$\frac{\text{number of ways an event can occur}}{\text{number of ways an event cannot occur}}$$

The **odds against** an event are found using this ratio:

$$\frac{\text{number of ways an event cannot occur}}{\text{number of ways an event can occur}}$$

Lisa has shot 50 free throws. She has made 35 and missed 15. From this information, you can find the odds.

Odds in favor of making next free throw $= \frac{35}{15} = \frac{7}{3}$

▶ Odds of $\frac{7}{3}$ can also be written as 7:3. Also, $\frac{7}{3}$ is a ratio and is always read as "seven to three."

Odds against making next free throw $= \frac{15}{35} = \frac{3}{7}$

▶ You can write the odds of $\frac{3}{7}$ as 3:7. Always read these odds as "three to seven."

Lisa interprets odds of 7:3 like this, "For every 7 free throws I make, I will probably miss 3." She interprets 3:7 as, "For every 3 free throws I miss, I will probably make 7."

Try THESE

Solve.

1. A coin is tossed.
   a. Find P(heads).
   b. Find P(not heads).
   c. Find the odds in favor of heads.
   d. Find the odds against heads.

2. A number cube is rolled.
   a. Find P(5 or 6).
   b. Find P(neither 5 nor 6).
   c. Find the odds in favor of a 5 or a 6.
   d. Find the odds against either a 5 or a 6.

Exercises

Solve.

1. A number cube is rolled. Find the odds against rolling a number less than 4.

2. If you guess the answer, what are the odds of getting a true-false test question correct?

3. A multiple-choice test question has 4 choices for each answer. What are the odds of guessing the correct answer?

4. Toss a coin 100 times. Keep a record of the outcomes. Based on this experiment, what are the odds in favor of heads?

Problem SOLVING

Find the odds for spinning each of the following.

5. for a country in Europe
6. in favor of Brazil
7. against a country in Europe
8. in favor of a country in Africa or a country in South America
9. in favor of a country in Asia or in North America

Solve.

10. Explain the expression "the odds are even."
11. Explain the difference between the odds of 1 to 2 and 2 to 1.

Mixed REVIEW

Compute.

12. $1\frac{1}{2} + 3\frac{4}{5}$

13. $-4\frac{5}{7} + -3\frac{1}{5}$

14. $9\frac{1}{3} - 5\frac{3}{4}$

15. $-6\frac{2}{5} \cdot 3\frac{3}{4}$

16. $6\frac{3}{5} \div 2\frac{1}{3}$

17. $-8\frac{3}{8} \cdot -5\frac{2}{5}$

9.9 Problem-Solving Strategy: Acting It Out

Objective: to solve a problem by acting it out

Susan Lee has six U.S. coins equaling $1.15. However, she cannot make change for a dollar, a half-dollar, a quarter, a dime, or a nickel. Which six coins does Susan have?

Susan has six coins equaling $1.15, but she cannot make change. Find which coins she has.

Gather materials, and act out the problem. Use several coins ranging from pennies to half-dollars. Try different combinations of six coins that make $1.15 until you find a group that does not include change for $1, 50¢, 25¢, 10¢, or 5¢.

You could make change for a dollar, a half-dollar, or a dime.

You could make change for a dollar or a half-dollar.

This group solves the problem. There are four dimes, one quarter, and one half-dollar.

Does this group satisfy all the requirements?

1. There are six coins in the group.
2. The coins have a value of $1.15.
3. You cannot make change for a dollar, half-dollar, quarter, or nickel.

The solution is correct. Susan has four dimes, one quarter, and one half-dollar.

Solve. Act it out.

1. Carlos places 10 pennies in a horizontal line on a desk, 5 pennies in a vertical line, 8 pennies in another vertical line, and 5 pennies in a diagonal line. What is the least number of pennies he can use?

2. Anne Zody buys a dog for $100, sells it for $110, buys it back for $120, and sells it again for $130. How much money does Ms. Zody make or lose?

Solve

1. Two children weighing 95 pounds each and their mother weighing 130 pounds want to cross a pond. The boat they have is safe for 280 pounds at a time. Using only the boat, how can they all get across the pond?

2. Martin had 40 baseball cards. He traded 7 of his cards for 5 of Dani's. He traded 3 more for 4 of Ted's. He traded another 2 for 1 of Anita's. He traded 11 more for 8 of Doug's. How many baseball cards does Martin have now?

3. Mrs. Vasquez and Mrs. Cochran bowl together. Their combined score for the first game was 262. Mrs. Vasquez's score was 40 higher than Mrs. Cochran's. What was each woman's score?

4. Carla is challenged to find her way through a maze. Each time she walks through a correct doorway, she receives $1. Each correct room she enters, she is given a reward equal to the amount she has already earned. On the correct path, Carla will pass through 8 doorways and 7 rooms. If she chooses the correct path, how much money will she receive?

★ 5. Mr. Howe sold 100 shares of stock for $5,975. When the price per share went down $5, he bought 200 shares, and he sold 100 of them when the price per share went back up $3. How much did Mr. Howe gain or lose in his transactions?

Mind BUILDER

Logical Thinking III

Four disks are stacked on a post as shown at the right. Without placing any disk over a smaller one, move the tower, one disk at a time, to one of the other posts. Find the *minimum* number of moves. Try to discover the pattern between the number of disks and the number of moves.

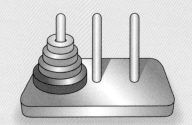

Chapter 9 Review

Language and Concepts

Choose the letter of the correct term or number to complete each sentence.

1. A(n) _____ is a list of all possible outcomes.
2. The _____ of an event is a ratio.
3. _____ events do affect one another.
4. A(n) _____ is one method of listing a sample space.
5. A(n) _____ can consist of any part of a sample space.
6. An arrangement in which order is important is called a(n) _____.
7. _____ events do not affect one another.
8. Events that cannot happen at the same time are _____.
9. A(n) _____ is a grouping in which order is not important.
10. An impossible event has a probability of _____.
11. A certain event has a probability of _____.

a. 1
b. permutation
c. combination
d. tree diagram
e. dependent
f. sample space
g. 0
h. odds
i. event
j. mutually exclusive
k. independent

Skills and Problem Solving

Answer each question using the spinner at the right. Sections 9.1, 9.8

12. What is the sample space for this spinner?
13. What is the probability that the spinner stops on 5 or yellow?
14. What is the probability that the spinner stops on 9?
15. What is the probability that the spinner stops on a number less than 5 or blue?
16. What is the probability that the spinner stops on an even number? on an odd number?
17. What are the odds in favor of the spinner stopping on 4?
18. What are the odds against the spinner stopping on red?

Draw a tree diagram and use the fundamental counting principle to find the number of possible outcomes. Section 9.2

19. You can buy a small, medium, or large pizza with thick or thin crust and with pepperoni or peppers.

20. Two coins are tossed, and a number cube is rolled.

21. A school cafeteria offers three choices for an entrée, two choices for a dessert, and five choices for a drink.

State whether each example is a *permutation* or a *combination*. Then solve. Sections 9.3–9.4

22. How many ways can you pair up students in a class of 20?

23. How many ways can 4 students stand in line at a drinking fountain?

24. If 15 teams play each other in the first round of a basketball tournament, how many games are played?

The table to the right shows the number of times a certain number was rolled when a number cube was rolled 75 times. Section 9.6

Number	Occurrence
1	12
2	9
3	15
4	7
5	14
6	18

25. What is the probability that you will roll a 6?

26. What is the probability that you will roll a 2?

27. What is the probability that you will not roll a 4?

28. If you roll the number cube 125 times, how many times would you expect to roll a 3?

A basket has 5 green lollipops, 12 red lollipops, and 7 orange lollipops. When a lollipop is taken from the basket, it is not replaced. Find each probability. Section 9.7

29. P(red)

30. P(orange, red)

31. P(red, red)

32. P(green, red, orange)

Solve. Sections 9.5, 9.8–9.9

33. Draw the first six rows of Pascal's triangle.

34. A coin is tossed. Find the odds in favor of heads.

35. Beverly and Pam are painting a fence. For every 3 sections Pam paints, Beverly paints 5. If Beverly paints 20 sections in 2 hours, how long will it take them to paint 18 sections?

Chapter 9 Test

The letters in the word *Paris* are written on slips of paper and placed in a hat. A slip of paper is drawn from the hat and replaced. Find each probability.

1. P(R)
2. P(Q)
3. P(not R)
4. P(R or a vowel)
5. P(R or a consonant)

The table shows the sample space of sums when two number cubes are rolled. Use the table to answer each question.

6. How many outcomes are in the sample space?
7. What is the probability of a sum of 7?
8. What is the probability of a sum less than 6?
9. What are the odds in favor of a sum of 7?
10. Suppose you roll two number cubes 1,000 times. How many sums greater than 8 would you expect?

Roll of Two Number Cubes

Sum	Frequency
2	1
3	2
4	3
5	4
6	5
7	6
8	5
9	4
10	3
11	2
12	1
Total	36

Solve.

11. Refer to the *Paris* letters above. What is the probability of drawing an **A**, not replacing it, and then drawing a vowel?

12. Three people are running for office. In how many ways can one be elected as president and one as treasurer?

13. How many 5-digit numbers can you write using the digits 2, 3, 4, 5, and 6? In each number you write, you can use each digit only once.

14. Four people are speaking at a program. How many ways can they be seated at the speakers' table?

15. Susan Parker plays on a softball team. Her statistics for the season are given in the table below. How many hits would you expect Susan to get in her next 50 plate appearances?

Hits	32
Walks	6
Outs	42
Plate Appearances	80

16. A candy bowl has 15 red candies, 21 green candies, 9 blue candies, and 11 yellow candies.

   a. What is the probability of picking a red candy, a blue candy, and then a red candy if the candies are replaced?

   b. What is the probability of picking a red candy, a blue candy, and then a red candy if the candies are not replaced?

   c. Explain why the answers to parts **a** and **b** are different.

Change of Pace

Is This Game Fair?

Play the following game several times with a partner. Use a marker (like a coin or other object) and the game board shown at the right.

Follow these rules.

- Player 1 puts the marker in any one of the empty squares in the top row.
- Player 2 moves the same marker one square to the right, one square to the left, or one square down. No player can move up or diagonally.
- Player 1 and Player 2 take turns.
- A player is not allowed to move to the previously occupied square.
- The first player who moves the marker into the winning area wins the game.

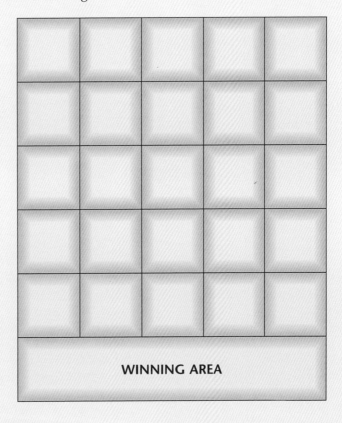

Solve.

1. Can you find a winning strategy?
2. Does a player appear to gain an advantage by playing first or by playing second?
3. With your playing partner, analyze the game. Try to find a winning strategy. That is, try to find a way in which you could win every time.
4. Is this game fair or unfair? Explain your answer.

Cumulative Test

1. What is the median of the scores of the class in problem 4?
   a. 15
   b. 15.3
   c. 16.5
   d. none of the above

2. What is the mode of the set of data below?
   3, 6, 7, 5, 5, 6, 2, 6
   a. 5
   b. 5.5
   c. 6
   d. There is no mode.

3. What is the range of the number of days in the months of a nonleap year?
   a. 3
   b. 12
   c. 28
   d. 30

4. What is the probability of spinning blue?

   a. $\frac{1}{4}$
   b. $\frac{1}{3}$
   c. $\frac{1}{2}$
   d. $\frac{3}{4}$

5. A six-sided number cube is rolled. What is the probability of rolling a number other than 2?
   a. $\frac{1}{6}$
   b. $\frac{2}{6}$
   c. $\frac{5}{6}$
   d. none of the above

6. In how many ways can 5 photographs be lined up in a photo album?
   a. 5
   b. 25
   c. 120
   d. 125

7. What is the median? Use the stem-and-leaf plot.
   a. 5
   b. 84
   c. 85
   d. 71

Stem	Leaf
7	1 1 2
8	2 5
9	0 4 5 6

   7|1 = 71

8. The table shows test scores for a class of 23 students. How many in the class made a score greater than 15?
   a. 5
   b. 10
   c. 13
   d. 16

Score	Tally	Frequency
20	II	2
18	IIII	5
17	III	3
15	IIII	5
14	II	2
13	I	1
12	III	3
10	II	2

9. Name the polygon by the number of sides.
   a. decagon
   b. hexagon
   c. octagon
   d. pentagon

   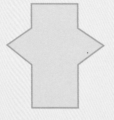

10. A box contains 3 red balls, 2 blue balls, and 1 white ball. If you pick one ball without looking, what is the probability that it will be either blue or red?
    a. $\frac{1}{6}$
    b. $\frac{1}{2}$
    c. $\frac{2}{3}$
    d. $\frac{5}{6}$

CHAPTER 10

More Equations and Inequalities

Rachel Reiley
Hawaii

10.1 Writing Two-Step Equations

Objective: to write and solve two-step equations

A taxicab charges a $3 fee plus an additional $1.50 for every mile driven. Your cab ride cost $12. How many miles did you travel?

You can translate this problem into an equation and solve it.
Let m = miles traveled.

Taxicab fee plus mile fee equals total ride cost.
$$3 + 1.5m = 12$$

You can now solve the equation.
$$3 - 3 + 1.5m = 12 - 3$$
$$\frac{1.5m}{1.5} = \frac{9}{1.5}$$
$$m = 6$$

You traveled 6 miles.

Examples

Sentence	Equation
A. Six less than four times a number is two.	$4n - 6 = 2$
B. Nine is twice a number increased by one.	$9 = 2x + 1$
C. A number divided by 8 added to 5 is -3.	$\frac{n}{8} + 5 = -3$

D. Write an equation and solve.

An ice-cream sundae costs $3 plus $2 for each topping. If Rosie paid $7 for her sundae, how many toppings did she get on it?

Sundae plus $2 for each topping equals $7.
$$3 + 2t = 7$$

$3 + 2t = 7$	Write the equation.
$3 - 3 + 2t = 7 - 3$	Subtract 3 from each side.
$\frac{2t}{2} = \frac{4}{2}$	Simplify.
	Divide each side by 2.
$t = 2$	Simplify.

Rosie got 2 toppings on her sundae.

Try THESE

Translate each sentence into an equation. Then solve each equation.

1. Seven plus the quotient of some number and four is nine.
2. The sum of six and the product of four and some number is negative eighteen.
3. Three less than the product of negative four and some number is negative fourteen.
4. Eight less than the quotient of some number and three is negative sixteen.

Exercises

Solve each problem by writing an equation and then solving the equation.

1. Jay Hawkins sells his stereo for $235. This is $10 more than half of what he paid for it. How much did he pay for his stereo?
2. The local swim club has a pool whose length is 5 feet more than twice its width. What is the width of the pool if the length is 25 feet?
3. At the grocery store, Jane buys a box of cereal for $3 and bananas for $0.55 a pound. If Jane spends $5.20 at the grocery store, how many pounds of bananas does she buy?
4. Shannon Steiner opens a savings account with $50. Each week after that she deposits $8 into her account. How many weeks does it take her to save $450?
5. Meg buys some books from an Internet bookstore. Each book costs $12, and she is charged $5 for shipping. If her total bill is $77, how many books does she buy?

Constructed RESPONSE

6. Josh and Jessica each want to save $150 for their summer rafting trip. Josh earns $15 an hour tutoring. Jessica has $30 saved and earns $8 an hour baby-sitting.
   a. Write two equations to show how long it will take Josh and Jessica to each save $150.
   b. Who will take longer to save $150?
   c. Explain.

Test PREP

7. Ernesto loses 14.5 pounds in 7 months. Now he weighs 156.5 pounds. Which equation can be used to find how much he weighed before?
   a. $x + 14.5 = 156.5$
   b. $x - 14.5 = 156.5$
   c. $7x + 14.5 = 156.5$
   d. $7x - 14.5 = 156.5$

8. A rectangle has a length of 15 inches and a perimeter of 50 inches. What is the width of the rectangle?
   a. 30 inches
   b. 20 inches
   c. 10 inches
   d. 5 inches

10.2 Simplifying Algebraic Expressions

Objective: to simplify algebraic expressions

Lori is a reservations agent. Her job is to find connecting flights, so people can change planes without waiting too long at the airport. To do this, she must match arrival times and departure times as closely as possible.

To simplify algebraic expressions, you match **terms** and combine them. A term is a number, a variable, or the product of a number and a variable or variables. The expression $7 + 3x + 5x$ has three terms. Terms that have the same variables are called **like terms**.

Like Terms	Not Like Terms
6 and -15	$6m$ and -15
$-6x$ and $4x$	$-6x$ and $4y$
$5xy$ and $7xy$	$5x$ and $7xy$

Examples

The distributive property can be used to combine like terms.

A. $3x + 5x = (3 + 5)x$ distributive property
         $= 8x$ Add.

B. $5x + 4x + 6 = (5 + 4)x + 6$
                $= 9x + 6$

C. $m - 6m = 1m - 6m$ Rewrite m as $1m$.
           $= (1 - 6)m$ distributive property
           $= -5m$

You can also use the distributive property to rewrite expressions.

D. $6(y + 9)$
     $6(y + 9) = 6 \bullet y + 6 \bullet 9$ Distribute.
               $= 6y + 54$ Simplify.

E. $-2(g - 5) = -2(g + -5)$ Rewrite $g - 5$ as $g + -5$.
               $= -2 \bullet g + -2 \bullet -5$ Distribute.
               $= -2g + 10$ Simplify.

Try THESE

Use the distributive property to rewrite each expression.

1. $7(x + 8)$
2. $-8(h + 10)$
3. $2(m - 4)$
4. $-5(p + 6)$
5. $12(2t + 2)$
6. $-\frac{1}{2}(6a - 4)$

Exercises

Combine like terms.

1. $-2y + 6y$
2. $3w - 10w$
3. $-11r + 5r$
4. $-24z - 15z$
5. $7m + m$
6. $12g - 5g$

Simplify each expression.

7. $6x + 8x + 2$
8. $9a + 9a + 9b$
9. $2c - 4 + 8c + 8y$
10. $y + 1.2y + 1.2z$
11. $6r + r - 5r$
12. $5t + 5s + 5xs$
13. $\frac{3}{4} + 4x - 5x + 9$
14. $3x + 2 - 8 - 3x$
15. $5m - 6t + 4m + 3t$
16. $6d - \frac{1}{2}d + \frac{1}{2}$
17. $5 - 8n + n$
18. $m - 5 - 2m + 5$
19. $12 + 3x - 5x$
20. $7.9k + 4 - 3.5k + 2$
21. $2z - 3y - 7z + y$
22. $5x + 2(x + 6)$
23. $-3m + 3(m + 6)$
24. $2(x + 6) - 7x + 3$

Problem SOLVING

25. Write two different expressions that can be simplified to $2x + 5$.
26. Write an expression in simplest form for the perimeter of the triangle below.

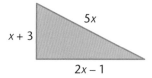

MIXED REVIEW

Evaluate each expression.

27. $12 + 3^2 - 5$
28. $7 \cdot 6 - 3$
29. $12 - 3 \cdot 4 - \sqrt{81}$

Solve each equation.

30. $x - 8 = 41$
31. $9 = m + 2$
32. $n - 1.4 = 1.7$

10.3 Solving Multi-Step Equations

Objective: to solve equations with multiple steps

Terry drove 5 hours on a trip and stopped for lunch. He then drove 4 more hours in the rain before reaching his destination. If the trip was 475 miles, what was his average speed?

Write an equation to represent Terry's trip. If s represents his speed, then the equation is $5s + 4s = 475$. Now solve the equation. Round the answer to the nearest tenth.

$$5s + 4s = 475$$
$$9s = 475 \quad \text{Combine like terms.}$$
$$\frac{9s}{9} = \frac{475}{9}$$
$$s = \frac{475}{9}$$
$$s = 52.\overline{7}$$

Terry's average speed was about 52.8 miles per hour.

Another Example

Solve for x.

$2(x + 3) - x = 15$
$2x + 6 - x = 15$ distributive property
$2x - x + 6 = 15$ commutative property
$x + 6 = 15$ Combine like terms.
$x + 6 - 6 = 15 - 6$
$x = 9$

Check: $2(x + 3) - x = 15$
$2(9 + 3) - 9 \stackrel{?}{=} 15$
$2(12) - 9 \stackrel{?}{=} 15$
$24 - 9 \stackrel{?}{=} 15$
$15 = 15$ ✓

Try THESE

Combine like terms to form a new equation.

1. $5x - 2x = 24$
2. $7y - 13y = 120$
3. $8n + 10n = 4$
4. $4.2 = 10m + 10.7 - 2m$
5. $\frac{2}{3}x - \frac{1}{3} + 2x = 37$
6. $r + 2r + 5r = 96$

Exercises

Solve each equation. Check your solution. *Show all Steps! cross products multiply in ( ) [ ]*

1. $5(y - 1) = 50$
2. $29 = -13 + 8n - n$
3. $\frac{2}{3}(r - 12) + \frac{4}{3}r = 26$
4. $12 = 2(x - 3) + 7x$
5. $(2 - w) + (3 - 2w) = 32$
6. $4(x - 3) - 3x = 0$
7. $225 = 17y - 2(15 + y)$
8. $1.5 - 4x + 16x = 3.9$
9. $\frac{x}{3} + \frac{2x}{3} - 1 = 7$
10. $42 = 3(7 - x)$
11. $3.1n + 4.9n - 15 = 65$
12. $210y - 66y + 72 = 228$

Problem SOLVING

13. Armando rides his bike to the store in 8 minutes. He rides home in 10 minutes. His home is 9 blocks from the store. What is his average time to ride 1 block?

★14. If a number is doubled and added to a number 1 more than the original number, the result is 25. Find the number.

15. If twice a number is subtracted from 8, the difference is 12. What is the number?

16. Three consecutive integers can be represented by n, $n + 1$, and $n + 2$. Their sum is 66. Find the numbers.

17. **Make Up a Problem**
    Write a problem using the following information.
    w = Noell's weight in kilograms
    $w - 4 = 39$

Mid-Chapter REVIEW

Solve each equation. Write each solution in simplest form.

1. $h - \frac{7}{2} = 2\frac{1}{2}$
2. $\frac{x}{8} = -7.3$
3. $-9s = 7.2$
4. $\frac{7x}{8} = 21$
5. $-6q + 5.8 = 23.8$
6. $\frac{b - 5}{-2} = 8$

Simplify.

7. $2(c + 2d) + 5(2c + 3d)$
8. $19pq + 5p + 3pq + 7p - 12p$

Write an equation, and then solve the equation.

9. The sum of 5 and 4 times a number is -11. Find the number.
10. The product of 3 and the difference of a number and 10 is 15. Find the number.

10.4 Solving Equations with Variables on Both Sides

Objective: to solve equations with variables on both sides

On a wind gauge, the wind velocity measured during a strong gust is 2 mph less than 3 times the steady wind velocity. The gust is also 16 mph greater than the steady rate. What is the steady wind velocity?

Let v represent the velocity of the steady wind. Use the equation $3v - 2 = v + 16$ to solve this problem.

$$3v - 2 = v + 16$$
$$3v + {-v} - 2 = v + {-v} + 16 \quad \text{Add the opposite of } v \text{ to both sides.}$$
$$2v - 2 = 16$$
$$2v - 2 + 2 = 16 + 2 \quad \text{Add 2 to both sides.}$$
$$2v = 18$$
$$\frac{2v}{2} = \frac{18}{2} \quad \text{Divide both sides by 2.}$$
$$v = 9$$

Check:
$$3v - 2 = v + 16$$
$$3 \cdot 9 - 2 \stackrel{?}{=} 9 + 16$$
$$27 - 2 \stackrel{?}{=} 25$$
$$25 = 25 \checkmark$$

The velocity of the steady wind is 9 mph.

Another Example

Solve for y.

$$8y - 3 = 1 + 10y$$
$$8y + {-8y} - 3 = 1 + 10y + {-8y} \quad \text{Add the opposite of } 8y \text{ to both sides.}$$
$$-3 = 1 + 2y$$
$$-3 + {-1} = {-1} + 2y \quad \text{Add the opposite of 1 to both sides.}$$
$$-4 = 2y$$
$$\frac{-4}{2} = \frac{2y}{2} \quad \text{Divide both sides by 2.}$$
$$-2 = y$$

Check:
$$8y - 3 = 1 + 10y$$
$$8 \cdot {-2} - 3 \stackrel{?}{=} 1 + 10 \cdot {-2}$$
$$-16 - 3 \stackrel{?}{=} 1 + {-20}$$
$$-19 = -19 \checkmark$$

Try THESE

Solve and check each equation.

1. $2x = 15 - x$
2. $2x + 12 = 18 - x$
3. $8x + 3 = 15x + 18$
4. $2 - n = 7 - 2n$
5. $\frac{1}{2}p + 6 = \frac{1}{3}p + 7$

Exercises

Solve each equation.

1. $7r - 7 = 2r + 18$
2. $1.5h + 7.2 = 5.2 + 3h$
3. $8x + 15 = 4x - 5$
4. $6x - 6 = 4x + 16$
5. $3y - 19 = 7y - 55$
6. $0.2s = 0.7s - 1.4$
7. $4x + 4 = 9x - 36$
8. $-3y + 5 = 2y - 20$
9. $\frac{1}{3}k - 2 = \frac{1}{6}k + 5$
10. $-4x + 14 = 2x + 2$
11. $8c + 5 = 3c$
12. $10x - 5 = 24x - 32$
13. $0.3n - 2.8 = 0.9n - 1$
14. $\frac{t}{2} = 3t + 5$
15. $\frac{x}{2} + 5 = \frac{3}{2}x + 1$
16. $\frac{z - 5}{-3} = 9 + 3.4z$

Problem SOLVING

17. Mike flew 2.5 hours and stopped for fuel. After a half hour of rest, he flew another 3 hours to his destination. The total trip was 660 miles. What was Mike's average flying speed?

18. On Monday, the barometric pressure was 0.08 greater than on Sunday. On Tuesday, the pressure was 0.10 less than on Sunday. If Monday's reading was 30.06, what was the pressure on Tuesday?

19. An airline ticket from Dallas to Los Angeles costs $100 more than twice what it cost 10 years ago. If the price now is $500, what was the price 10 years ago?

★20. Find three consecutive even integers such that 3 times the least is 4 more than twice the greatest.

21. Two ways of representing the same number are $5n - 3$ and $4n + 3$. Find the value of n and the number.

Problem Solving

The Container Problem

Jack and Jill went up a hill to fetch a liter of water. Jack has a 2-liter pail. Jill has a 5-liter pail.

1. How can they measure out exactly 1 liter of water?

2. Which of the following amounts could Jack and Jill measure?

   1 liter   2 liters   3 liters   4 liters

   5 liters   6 liters   7 liters   8 liters

Extensions

1. What amounts could Jack and Jill measure using only a 3-liter and an 11-liter pail?

2. Name three containers that will measure from 1 to 20 liters using no other containers.

Cumulative Review

Evaluate each expression if $a = 3$, $b = 4$, and $c = -6$.

1. $4a$
2. $c - 1$
3. $c^2 - 4$
4. $a - (b - c)$
5. $(a - b)(b + 3c)$
6. $2a^2 + bc$

Write each fraction as a decimal. Use bar notation for repeating decimals.

7. $\frac{9}{8}$
8. $\frac{7}{50}$
9. $\frac{5}{16}$
10. $\frac{8}{9}$
11. $\frac{1}{6}$

Write each decimal as a fraction in simplest form.

12. 0.19
13. 2.7
14. $0.\overline{5}$
15. 1.06
16. 0.043

Solve each equation. Check your solution.

17. $x + 3 = 8$
18. $3b = 16$
19. $2y - 5 = 27$
20. $\frac{c}{9} + 1 = 2.3$

Solve each proportion.

21. $\frac{3}{4} = \frac{c}{10}$
22. $\frac{4.6}{6.9} = \frac{14}{a}$
23. $\frac{6.4}{4.8} = \frac{b}{3.6}$
24. $\frac{5.1}{3.4} = \frac{81}{k}$

Write each fraction or decimal as a percent.

25. $\frac{3}{4}$
26. $2\frac{1}{2}$
27. $\frac{4}{5}$
28. $\frac{9}{8}$
29. $\frac{4}{9}$
30. 0.46
31. 0.07
32. 3.16
33. 0.542
34. 0.008

Solve.

35. What number is 35% of 72?

36. What percent of 68 is 17?

37. Justin is traveling from Cleveland to Cincinnati. He has a choice of plane, rail, or bus transportation; day or night trips; and express (nonstop) or local travel. How many possible ways can Justin make the trip?

38. Pizza can be ordered with thick or thin crust, with regular or extra cheese, and with five different toppings. How many different kinds of pizza can be ordered?

39. Membership in the City Zoo is $12.50 per person. If there are more than four in a family, the zoo offers a 50% discount for each additional member. How much will the Álvarez family pay for six members?

40. Janice sees a record on sale at Discland for 10% off the regular price of $8.30. The Turntable has the same record for $8.95 with a 15% discount. At which store is the record less expensive and by how much?

The following set of data represents inches of snowfall in several cities during March.

6, 14, 19, 27, 36, 42, 63, 87

41. Find the mode.
42. Find the mean.
43. Find the median.

10.5 Solving Inequalities by Addition and Subtraction

Objective: to solve inequalities by addition and subtraction and to graph the solutions

An **inequality** is a mathematical sentence comparing two quantities using symbols, such as <, >, ≤, ≥, or ≠. The set of all replacements for the variable that makes the inequality true is called the **solution set** of the sentence. An inequality can be solved like an equation, and the solution can be graphed on a number line.

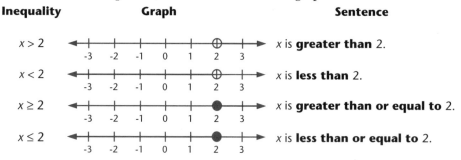

Inequality	Graph	Sentence
$x > 2$		x is **greater than** 2.
$x < 2$		x is **less than** 2.
$x \geq 2$		x is **greater than or equal to** 2.
$x \leq 2$		x is **less than or equal to** 2.

When graphing inequalities, an **open circle** indicates the number is not included, meaning greater than or less than, and a **closed circle** indicates the number is included, meaning greater than or equal to or less than or equal to. An arrow to the left or right indicates the direction of the inequality.

Addition and Subtraction of Inequalities

If you add or subtract the same number from each side of an inequality, the inequality remains true.

If $a > b$, then $a + c > b + c$ and $a - c > b - c$.
If $a < b$, then $a + c < b + c$ and $a - c < b - c$.

$7 > -2$	$12 < 20$
$7 + 3 > -2 + 3$	$12 - 6 < 20 - 6$
$10 > 1$	$6 < 14$

You can solve inequalities the same way you solve equations.

Example

Solve. $n + 8 < 6$

$n + 8 < 6$

$n + 8 - 8 < 6 - 8$ Subtract 8 from both sides.

$n < -2$ Simplify.

Check: $n + 8 < 6$

$-3 + 8 < 6$

$5 < 6$

The solution is $n < -2$.

Graph the solution.

Place an open circle at -2.
Draw a line and an arrow to the left.

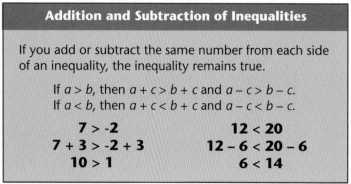

Try THESE

Graph each inequality on a number line.

1. $x < -3$
2. $m \leq -3$
3. $y < 6$
4. $x \geq -1.5$

Exercises

Solve each inequality. Check your solution.

1. $m - 7 < 6$
2. $n - 8 > 5$
3. $x + 4.5 \geq 5.5$
4. $r - 9 \leq 7$
5. $p + 12 \geq 9$
6. $4 + x < -2$
7. $a - 6 > 2.5$
8. $12 + b \geq -5$

Translate each sentence into an inequality. Then solve each inequality.

9. Five less than a number is at least nine.
10. Four more than a number is greater than seven.
11. The difference between a number and three is less than fifteen.

Solve each inequality. Check your solution. Then graph the solution on a number line.

12. $p + 5 < 10$
13. $m - 4.5 \geq 6$
14. $h - 2.8 \leq -4$
15. $a + 2\frac{1}{2} > 3$
16. $g - \frac{2}{3} < \frac{1}{2}$
17. $a - 3 < -5$
18. $2 \geq w + 9$
19. $8 + p < -6$

Problem SOLVING

20. Raul is going on a trip and wants to take at least $80 with him. He currently has $25. Write and solve an inequality to show how much money Raul needs to withdraw from the bank.

21. Julia is allowed to watch no more than 5 hours of television a week. So far this week she has watched 1.5 hours. Write and solve an inequality to show how many hours of television Julia can still watch this week.

10.5 Solving Inequalities by Addition and Subtraction

10.6 Solving Inequalities by Multiplication and Division

Objective: to solve inequalities by multiplication and division

Exploration Exercise

1. Start with the inequality $15 > 9$.
2. Multiply both sides of $15 > 9$ by 3. What do you notice about the inequality?
3. Divide both sides of $15 > 9$ by 3. What do you notice?
4. Multiply both sides of $15 > 9$ by -3. What do you notice? $15 > 9$ by -3 true
5. Divide both sides of $15 > 9$ by -3. What do you notice? $-5 < -3$

The above exploration demonstrates the multiplication and division properties of inequalities.

Multiplying or Dividing an Inequality by a Positive Number

When you multiply or divide both sides of an inequality by the same positive number, the inequality remains true.

If $a > b$, then $ac > bc$ and $\frac{a}{c} > \frac{b}{c}$ with $c > 0$. If $a < b$, then $ac < bc$ and $\frac{a}{c} < \frac{b}{c}$ with $c > 0$.

$$7 > -4$$
$$3 \cdot 7 > 3 \cdot -4$$
$$21 > -12$$

$$12 < 14$$
$$\frac{12}{2} < \frac{14}{2}$$
$$6 < 7$$

Examples

A. Solve the inequality $2p \leq 14$.

$2p \leq 14$ Check: $2p \leq 14$

$\frac{2p}{2} \leq \frac{14}{2}$ Divide both sides by 2. $2 \cdot 6 \leq 14$

$p \leq 7$ Simplify. $12 \leq 14$

The solution is $p \leq 7$.

In parts 4 and 5 of the exploration above, you saw that you needed to reverse the direction of the inequality sign when you multiplied and divided both sides of the inequality by a negative number to make the inequality true.

Multiplying or Dividing an Inequality by a Negative Number

When you multiply or divide both sides of an inequality by the same negative number, you must reverse the direction of the inequality symbol for the inequality to remain true.

If $a > b$, then $ac < bc$ and $\frac{a}{c} < \frac{b}{c}$ with $c < 0$. If $a < b$, then $ac > bc$ and $\frac{a}{c} > \frac{b}{c}$ with $c < 0$.

$7 > -4$
$-3 \cdot 7 < -3 \cdot -4$
$-21 < 12$

$12 < 14$
$\frac{12}{-2} > \frac{14}{-2}$
$-6 > -7$

B. Solve the inequality $\frac{x}{-5} > 7$.

$\frac{x}{-5} > 7$

$-5 \cdot \frac{x}{-5} < 7 \cdot -5$ Multiply both sides by -5 and reverse the inequality sign.

$x < -35$ Simplify.

The solution is $x < -35$.

Check: $\frac{x}{-5} > 7$

$\frac{-40}{-5} > 7$

$8 > 7$

Try THESE

State whether you should reverse the direction of the inequality sign when solving. Write *yes* or *no*.

1. $5x > 25$
2. $\frac{x}{-9} \leq 2$
3. $8x < -64$
4. $-6m < 36$

Exercises

Solve each inequality. Check your solution. Then graph the solution on a number line.

1. $9m < 45$
2. $6x \leq -36$
3. $\frac{b}{3} > 2$
4. $-10c > -10$
5. $\frac{k}{-6} < 14$
6. $-x > 12$
7. $-50 \leq 25p$
8. $\frac{y}{1.2} \geq 4$
9. $\frac{m}{-3} \geq -15$
10. $-7 < \frac{h}{9}$
11. $1.6h \leq -64$
12. $-8 \geq \frac{x}{-0.4}$

Problem SOLVING

13. A movie theater charges $7.50 to see a movie. They also offer a yearly pass for $90. Write and solve an inequality to find out how many movies you would need to see so that the yearly pass is cheaper than paying for individual movies.

10.7 Solving Two-Step Inequalities

Objective: to solve two-step inequalities

You are baking and have a bag of sugar that contains at least 7 cups. You are making a cake that requires 2 cups of sugar, and you want to make cookies where each batch requires 2.5 cups of sugar. How many batches of cookies can you make?

You can solve the inequality $2.5b + 2 \leq 7$ to find the number of batches of cookies you can make. This is a two-step inequality. Solving a two-step inequality is similar to solving a two-step equation.

When solving a two-step inequality, you must complete the following steps:
1. Undo the addition or subtraction.
2. Undo the multiplication or division.

✳ Remember to reverse the direction of the inequality sign when you multiply or divide both sides of an inequality by a negative number.

Examples

A. Solve the inequality $2.5b + 2 \leq 7$.

		Check: $2.5b + 2 \leq 7$
$2.5b + 2 \leq 7$		
$2.5b + 2 - 2 \leq 7 - 2$	Subtract 2 from each side.	$2.5 \cdot 2 + 2 \leq 7$
$2.5b \leq 5$	Simplify.	$5 + 2 \leq 7$
$\dfrac{2.5b}{2.5} \leq \dfrac{5}{2.5}$	Divide each side by 2.5.	$7 \leq 7$
$b \leq 2$	Simplify.	

B. Solve the inequality $-3t + 2 > 14$.

$-3t + 2 > 14$

$-3t + 2 - 2 > 14 - 2$ Subtract 2 from each side.

$-3t > 12$ Simplify.

→ $\dfrac{-3t}{-3} < \dfrac{12}{-3}$ Divide each side by -3. Reverse the direction of the inequality sign.

$t < -4$ Simplify.

You need to check your solution. Choose a number that is less than -4, and substitute it into the original inequality.

$-3 \cdot -5 + 2 > 14$

$15 + 2 > 14$ Replace t with -5.

$17 > 14$

Try THESE

Solve each inequality. Check your solution.

1. $2g + 8 > 12$
2. $\frac{x}{3} - 4 \leq 2$
3. $6d - 14 \geq 4$

Exercises

Solve each inequality. Check your solution.

1. $2m + 9 > 5$
2. $-5b - 15 \leq 20$
3. $\frac{a}{4} + 7 > 3$
4. $6 + 3g \geq 15$
5. $-4 + 2h \leq 14$
6. $-9x - 16 < 20$
7. $5 - \frac{x}{9} < 4$
8. $\frac{x}{7} + 2 \leq -5$
9. $20 \geq 4n + 16$
10. $8x - 4.2 < 12.6$
11. $\frac{p}{12} + 14 \leq -6$
12. $-9 + 8d \geq 23$
13. $5 \geq -4s + 9$
14. $4x + 12.3 > 21.1$
15. $-8 > -3x + 7$

Problem SOLVING

16. You are running a lemonade stand. You spent $5 on supplies, and you earn $1.50 for each cup of lemonade you sell. How many cups of lemonade do you need to sell to earn at least $25? Write an inequality, and then solve.

17. For a workout today, you want to stretch for 5 minutes, and then run. It takes you 9 minutes to run a mile. How many miles can you run if you have no more than 32 minutes to work out? Write an inequality, and then solve.

Constructed RESPONSE

18. Explain the similarities and differences between solving $-5x + 6 = 21$ and $-5x + 6 < 21$.

Test PREP

19. Which inequality can be used to solve the following problem: Danielle can spend at most $52 on pants and shirts. She buys a pair of pants for $16. She wants to buy shirts that are on sale for $12 each. How many shirts can she buy?

   a. $16s + 12 \leq 52$
   b. $12s + 16 \leq 52$
   c. $16s + 12 \geq 52$
   d. $12s + 16 \geq 52$

20. Which graph below is the solution to the inequality $-3x - 5 < -8$?

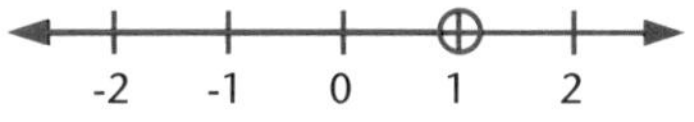

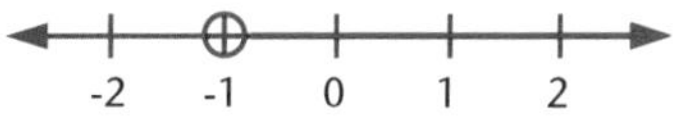

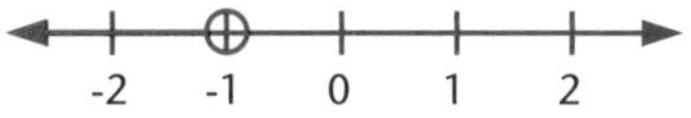

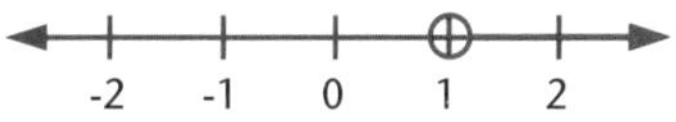

10.8 Problem-Solving Application: Using Inequalities

Objective: to solve problems using inequalities

Juanita is going on vacation. She plans to spend at most $60 for jeans and shirts to wear on the trip. She bought two shirts for a total of $24. How much can she spend on jeans?

Problems containing phrases, like *at most* and *at least*, can often be solved by using inequalities. Other phrases that suggest inequalities are *less than* and *greater than*.

You need to find out how much Juanita can spend on jeans. You know she has already spent $24 on shirts.

Let x represent the amount Juanita can spend on jeans. The total amount she can spend is equal to the amount she spends on shirts plus the amount she spends on jeans.

Total spent is less than or equal to $60.

$$24 + x \leq 60$$
$$24 - 24 + x \leq 60 - 24$$
$$x \leq 36$$

Juanita can spend at most $36 on jeans for her trip.

If Juanita spends more than $36, then the total will be more than $60. The answer is correct.

Write an inequality that describes each statement.

1. The number x is at least $\frac{1}{8}$.
2. The number y is at most 4.
3. Five times the number z is less than seven.
4. Ten times the number a is more than twelve.
5. The number x plus 1.2 is at least 4.5.
6. The number t minus 9 is at most 67.

Write an inequality that describes each problem.

1. Four times a number increased by four is at least sixteen. What is the number?

2. The sum of two consecutive positive odd integers is at most 20. What are the integers?

3. *The Columbus Dispatch* pays 5¢ per newspaper to the delivery person. How many newspapers must this person deliver to earn at least $3.50 per day?

4. George planned to spend at most $48 for shirts and ties. He bought 2 shirts for $18 each. How much can he spend for ties?

Use an inequality to solve each problem.

5. Five times a number increased by twelve is at least thirty-seven. What is the number?

6. Seven times a number decreased by four times the number is less than thirty. What is the number?

7. The sum of two consecutive positive odd integers is at most 18. What are the integers?

8. The sum of two consecutive positive even integers is at most 22. What are the integers?

9. A stove and a freezer weigh at least 260 kg. The stove weighs 115 kg. What is the weight of the freezer?

10. Ellen and her father spent at least $110.00 while shopping. Ellen spent $47.32. How much did her father spend?

Use the table to solve each problem.

11. Estimate which resort has the highest percent of discount from winter to summer.

12. Estimate the percent of discount that would be the highest among the resorts listed.

★13. If s = summer rate and w = winter rate, write an equation for the summer rate at Badger Run Resort.

Lodging	Winter Rates	Summer Rates
Riverton Mountain Resort	$140	$75
Timber Trail Manor	175	75
Stoney Creek Resort	135	75
Beaver Creek Inn	190	100
White Knoll Condos	164	89
Holly Gibson Lodge	129	89
Breakneck Village	115	70
Timberwolf Lodge & Condos	150	70
Brown Ridge Resort	154	61
Country Cottage House	84	50
Badger Run Resort	125	75

Chapter 10 Review

Language and Concepts

Choose a term from the list at the right to complete each sentence.

1. 8*ab* and 4.2*b* are examples of _____.
2. A mathematical sentence that contains a symbol, such as <, >, ≤, ≥, or ≠, is called a(n) _____.
3. A mathematical sentence with an equals sign is called a(n) _____.
4. Terms that have the same variables are called _____.
5. A value that makes an equation true is called a(n) _____.
6. The set of all replacements for the variable that makes an inequality true is a(n) _____.

solution set
terms
inequality
solution
inverse operations
equation
like terms

Skills and Problem Solving

Simplify. Section 10.2

7. $10x + x$
8. $\frac{3}{4}a + \frac{2}{3}ab + ab$
9. $9(r + s) - 2.8s$

Solve each equation. Sections 10.3–10.4

10. $3.5 = 1.7(w - 1) + 1$
11. $6(14 - 3x) + 20x = 90$
12. $\frac{1}{2}p - 7p + 3 = \frac{1}{8}$
13. $\frac{2}{3}x + 5 = \frac{1}{2}x + 4$
14. $8y - 4 = -10y - 50$
15. $2.9n + 1.7 = 3.5 + 2.3n$

Solve each inequality. Check your solution. Then graph the solution on a number line. Sections 10.5–10.7

16. $x - 6 < 15$
17. $-3d < 81$
18. $3 + x \geq 4$
19. $2y < 12$
20. $2n + 5 \leq 7$
21. $2 > n - 3$
22. $3r + 7 \leq -5$
23. $-4p + 5 < 21$
24. $\frac{x}{-3} + 6 \geq 4$

Write an equation, and then solve. Section 10.1

25. Dave bought a season ticket package to see his favorite baseball team. He paid $180, which included a processing fee of $40 and tickets to 15 games. How much does each game ticket cost?

26. At the sporting goods store, Kim bought a softball bat for $18 and 5 balls. She spent a total of $33. How much did each ball cost?

Use an inequality to solve each problem. Section 10.8

27. Two times a number increased by thirteen is at least three times that same number. What is the number?

28. If 5.3 times an integer is added to 20, the result is between 46.5 and 99.5. What is the integer?

29. Kara is planning a party. The cost to rent the space for the party is $50. The cost for the food is $15 per person. If Kara has at most $230 to spend on the party, how many people can she invite? Show all work, and explain your reasoning.

Chapter 10 Test

Write an equation, and then solve.

1. Charlie and three friends went to dinner. Each person ordered their own entrée which cost $8, but they all split an appetizer. If each person paid $12, how much did the appetizer cost?

Simplify.

2. $2.5x - x + y + 3.5y$

3. $3(a + 2) + 5a$

4. $4an + \frac{2}{3}am + 8an + \frac{1}{3}am$

Solve each equation.

5. $3x + 2 = 16$

6. $8r - \frac{r}{3} = 46$

7. $\frac{3}{4}n - \frac{2}{3}n = 5$

8. $40 - 10b = 4b$

9. $3y + 9 = 4y - 9$

10. $5.2a + 0.7 = 2.8 + 2.2a$

Solve each inequality. Check your solution. Then graph your solution on a number line.

11. $4y > -36$

12. $m - 8 > -15$

13. $-3f < 45$

14. $6 + -5x \geq 21$

15. $\frac{x}{7} + 3 \leq -5$

16. $3d - 8 \geq 16$

Use an inequality to solve each problem.

17. If eight times a number is decreased by three, then the result is between five and fifteen. What is the result?

18. Linda plans to spend at most $85 on jeans and shirts. She buys 2 shirts for $13.50 each. How much can she spend on jeans?

Change of Pace

Computers: Binary System

 Most computers use a **base-two** or **binary** numeration system. In this system, powers of 2 determine the place values. The only digits are 0 and 1. In a computer, 0 represents *no* or a switch in the *off* position. The digit 1 represents *yes* or a switch in the *on* position.

 A light off means 0.

 A light on means 1.

The place-value chart shows the value of each digit of the base-two numeral 101101_{two}. Read this numeral as "one zero one one zero one base two."

one hundred twenty-eights	sixty-fours	thirty-twos	sixteens	eights	fours	twos	ones
128	64	32	16	8	4	2	1
2^7	2^6	2^5	2^4	2^3	2^2	2^1	2^0
		1	0	1	1	0	1

Find the value of 101101_{two} as follows.

$101101_{two} = (1 \times 2^5) + (0 \times 2^4) + (1 \times 2^3) + (1 \times 2^2) + (0 \times 2^1) + (1 \times 2^0)$

$\phantom{101101_{two}} = (1 \times 32) + (0 \times 16) + (1 \times 8) + (1 \times 4) + (0 \times 2) + (1 \times 1)$

$\phantom{101101_{two}} = 32 + 0 + 8 + 4 + 0 + 1$

$\phantom{101101_{two}} = 45_{ten}$

Change 20 to a base-two numeral.

$20_{ten} = 10100_{two}$

> The powers of 2 are 1, 2, 4, 8, 16, and so on. 16 is the greatest power of 2 in 20. There is 1 sixteen with 4 remaining. In 4, there are 0 eights, 1 four, 0 twos, and 0 ones.

Find the value of each of the following.

1. 1_{two}
2. 10_{two}
3. 111_{two}
4. 1010_{two}
5. 1000_{two}
6. 10011_{two}
7. 100111_{two}
8. 10100010_{two}

Write each of the following as a base-two numeral.

9. 3
10. 4
11. 11
12. 14
13. 20
14. 32
15. 68
16. 224
17. the number of months in one year
18. your age in years
19. the number of days in the month of August
★20. the boiling point of water in degrees Celsius

Cumulative Test

1. Which is true?
   a. $-9 \cdot 8 > -1$
   b. $-9 + 8 < -1$
   c. $-9 + 8 \leq -1$
   d. none of the above

2. $80\% = $ _____
   a. 0.08
   b. 0.8
   c. 8
   d. 80

3. If there are 3 chairs and 3 different coverings, how many combinations are possible?
   a. 3
   b. 4
   c. 6
   d. 9

4. Which day has the highest temperature?

   Daily High Temperature
   Monday, -3°C
   Tuesday, -5°C
   Wednesday, 0°C
   Thursday, 2°C
   Friday, -1°C

   a. Monday
   b. Tuesday
   c. Wednesday
   d. Thursday

5. In the figure below, if Y is 6 miles east of X, how many miles is Z from Y?
   a. 6
   b. 12
   c. 36
   d. 270

6. Beth's times for running 5 km were 24 min, 22.5 min, 20.2 min, and 21.75 min. Which of the following is the first step in finding her average time?
   a. Add 24, 22.5, 20.2, 21.75, and 5.
   b. Add 24, 22.5, 20.2, and 21.75.
   c. Divide each time by 5.
   d. Divide 24 by 4.

7. Mr. Jhin bought a TV for $480. He paid 20% of the cost at the time of purchase. He paid the rest of the cost in 10 equal payments. How much was each payment?
   a. $9.60
   b. $38.40
   c. $48.00
   d. none of the above

8. A pentagonal-shaped building has hallways connecting each corner of the building to every other corner. How many of these hallways are there?
   a. 4
   b. 10
   c. 5
   d. 20

9. Using the information on the sign below, how much was the price reduced?

   a. 25%
   b. 33%
   c. 66%
   d. 75%

10. What statement(s) can be made from the following election results?

    Smith: 2,603   Jones: 1,460
    Yee: 3,092   Mendoza: 3,018

    a. Jones received the fewest votes.
    b. Yee received about 30% of the votes.
    c. No candidate received 50% of the votes.
    d. all of the above

280 Cumulative Test

CHAPTER 11

Linear Functions

Kristin O'Neill
Maryland

11.1 Sequences

Objective: to recognize and extend arithmetic and geometric series

Anna Rodriquez likes hiking in Shawnee Forest. On the first day, she hiked 2 miles. On the second day, she hiked 5 miles. On the third day, she hiked 8 miles. If she continues this pattern, how many miles will she hike on the fourth day?

The miles Anna hiked forms a **sequence**. A sequence is a list of numbers that follows a pattern. Each number in the list is a term. An **arithmetic sequence** is a sequence where the difference between any two consecutive terms is the same.

The difference is called the **common difference**.

A **geometric sequence** is a sequence where the quotient between any two consecutive terms is the same.

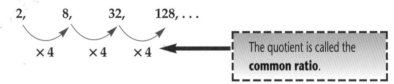

The quotient is called the **common ratio**.

Example

State whether the sequence 5, -15, 45, -135, . . . is *arithmetic*, *geometric*, or *neither*. Then find the next three terms.

5,   -15,   45,   -135, . . .
   ×-3   ×-3   ×-3

You can see that -15 ÷ 5 = -3, 45 ÷ -15 = -3, and so on.

The terms have a common ratio of -3, so the sequence is geometric. The next three terms are found by multiplying each consecutive term by -3.

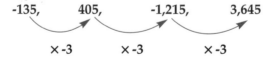

The next three terms are 405, -1,215, and 3,645.

Try THESE

Determine whether each sequence is *arithmetic*, *geometric*, or *neither*.

1. 10, 8, 6, 4, . . .
2. 1, 3, 4, 7, . . .
3. 6, -18, 54, -162, . . .

Exercises

Determine whether each sequence is *arithmetic, geometric,* or *neither* in problems 1–12. If it is arithmetic or geometric, state the common difference or ratio. Find the next three terms in each sequence.

1. 14, 19, 24, 29, . . .
2. 7, 10, 15, 22, . . .
3. 3, 6, 12, 24, . . .
4. -5, -1, 3, 7, . . .
5. 2, $3\frac{1}{2}$, 5, $6\frac{1}{2}$, . . .
6. -4, 8, -16, 32, . . .
7. -5, 10, -20, 40, . . .
8. 81, 27, 9, 3, . . .
9. 1, 4, 9, 16, . . .
10. 25, -5, 1, $-\frac{1}{5}$, . . .
11. 4, 6.5, 9, 11.5, . . .
12. 3, -3, 3, -3, . . .

13. What are the first four terms of an arithmetic sequence with a common difference of 1.5 if the first term is 15?

14. What are the first four terms of a geometric sequence with a common ratio of 10 if the first term is 4.5?

15. What are the first four terms of a geometric sequence with a common ratio of $-\frac{1}{2}$ if the first term is 40?

16. What are the first four terms of an arithmetic sequence with a common difference of -4 if the first term is 9?

Problem SOLVING

17. Write a sequence for the areas of the squares to the right. Is the sequence *arithmetic, geometric,* or *neither*?

18. Write a sequence for the perimeters of the squares to the right. Is the sequence *arithmetic, geometric,* or *neither*?

2 units   3 units   4 units

19. Write a sequence for the volume of cubes with side lengths of 2, 3, and 4. Is the sequence *arithmetic, geometric,* or *neither*?

Constructed RESPONSE

20. An ice-cream sundae at Susie's Sundaes costs $2.50 for three scoops of ice cream and $0.60 for each topping. Write a sequence of sundae prices with no toppings, one topping, two toppings, and three toppings. Is the sequence *arithmetic* or *geometric*? Explain how you know.

MixeD REVIEW

Solve each equation.

21. $2x - 3 = 7$
22. $7b - 4 = 2b + 16$
23. $\frac{x}{3} - 6 = 5$
24. $8x + 10 = 3x$
25. $8 = 2 + 3x$
26. $3x + 2 = 2x + 5$

11.2 Functions

Objective: to evaluate functions

Exploration Exercise

The Pep Team is selling brownies to raise money for new uniforms. They are selling brownies for $0.50 each.

Brownies Sold	Money Earned
1	$0.50
2	$1.00
3	
4	
5	

1. Copy and complete the table at the right.
2. If they sell 6 brownies, how much money will they earn?
3. Explain how you can find the amount of money earned if they sell 10 brownies.

The amount of money the Pep Team earns depends on, or is a function of, the number of brownies they sell. A **function** is a relationship where the output depends on the input. In a function, there is exactly one output value for every input value.

The amount of money the Pep Team earns can be modeled by the function $f(x) = 0.5x$.

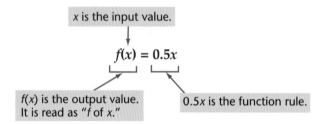

x is the input value.

$f(x) = 0.5x$

$f(x)$ is the output value. It is read as "f of x."

$0.5x$ is the function rule.

To find the value of a function, substitute a number into the function and simplify.

Examples

A. Find $f(3)$ if $f(x) = -2x + 5$.

$f(x) = -2x + 5$

$f(3) = -2 \cdot 3 + 5$ Substitute 3 for *x*.

$f(3) = -6 + 5 = -1$ Simplify.

Therefore, $f(3) = -1$.

You can also make a function table to find several outputs of a function.

B. Complete the function table.

Substitute each x-value into the function $x - 3$, and simplify to find the output.

Input x	Rule $x - 3$	Output $f(x)$
-2	-2 – 3	-5
-1	-1 – 3	-4
0	0 – 3	-3
1	1 – 3	-2
2	2 – 3	-1

The set of all the input values is the **domain**. The set of all the output values is the **range**. In Example B, the domain is {-2, -1, 0, 1, 2}, and the range is {-5, -4, -3, -2, -1}. In an ordered pair, (x, y), the x-values are the domain and the y-values are the range.

Try THESE

Find each function value.

1. $f(4)$ if $f(x) = x - 1$
2. $f(-7)$ if $f(x) = -8x$
3. $f(\frac{1}{2})$ if $f(x) = 4x$

Exercises

Find each function value.

1. $f(-9)$ if $f(x) = 5 - x$
2. $f(8)$ if $f(x) = 4 + 2x$
3. $f(\frac{3}{4})$ if $f(x) = 4x + 1$
4. $f(6)$ if $f(x) = -3x - 8$
5. $f(-5)$ if $f(x) = 5x - 25$
6. $f(4)$ if $f(x) = -x + 5$

Copy and complete each function table. State the domain and range of each function.

7. $f(x) = 2x + 2$

x	2x + 2	f(x)
-2		
-1		
0		
1		
2		

8. $f(x) = 7 - 3x$

x	7 – 3x	f(x)
-2		
-1		
0		
1		
2		

Problem SOLVING

9. One year in a dog's life is equivalent to about 7 years in a human's life. Felipe is 28 years old. What is the equivalent age for a dog?

11.3 Graphing Linear Functions

Objective: to graph linear functions

You can graph a function on a coordinate plane. The input values correspond to the **x-axis** (horizontal axis), and the output values correspond to the **y-axis** (vertical axis). You can make a table to find the input and output values, and then use the ordered pairs to graph the function.

Example

Graph the function $y = 3x - 1$.

Step 1
Choose several x-values. Make a function table.

x	3x – 1	y	(x, y)
0	3(0) – 1	-1	(0, -1)
1	3(1) – 1	2	(1, 2)
2	3(2) – 1	5	(2, 5)
3	3(3) – 1	8	(3, 8)

Step 2
Graph each ordered pair.

Step 3
Draw a line through the ordered pairs.

The coordinates of any point on the line will satisfy the equation. A function in which the graph of the solutions forms a line is called a **linear function**.

When you graph a line on a coordinate plane, you have some special points to look out for. The value of x, where the graph crosses the x-axis, is called the **x-intercept**. The value of y, where the graph crosses the y-axis, is called the **y-intercept**.

Another Example

Find the x- and y-intercepts of the function $y = 2x - 3$.

To find the x-intercept, set $y = 0$ and solve for x.

$0 = 2x - 3$ Let y = 0.
$0 + 3 = 2x - 3 + 3$ Add 3.
$3 = 2x$ Simplify.
$\frac{3}{2} = \frac{2x}{2}$ Divide by 2.
$\frac{3}{2} = x$ Simplify.

To find the y-intercept, set $x = 0$ and solve for y.

$y = 2 \cdot 0 - 3$ Let x = 0.
$y = 0 - 3$ Simplify.
$y = -3$ Simplify.

The x-intercept is $\frac{3}{2}$, and the y-intercept is -3.

Try THESE

Copy and complete each table. Then graph each equation.

1. $y = x$

x	y
0	0
1	■
2	■

2. $y = x - 4$

x	y
-1	-5
0	■
1	■

3. $y = 2x - 3$

x	y
-2	-7
0	■
2	■

4. $y = 7 - 3x$

x	y
-1	■
0	■
1	■

Exercises

Graph each function on graph paper. Find the x- and y- intercepts of each function.

1. $y = -1x$
2. $y = 3x$
3. $y = 2x + 1$
4. $y = 2x + 5$
5. $y = x - 5$
6. $y = 4 - x$
7. $y = 6 - 3x$
8. $y = \frac{1}{2}x - 5$

Problem SOLVING

9. The function $c = 2m + 10$ represents the cost c, in dollars, of renting a moving vehicle and driving m, miles from Mo's Moving. Graph the function. Use the graph to find the cost of renting a vehicle and driving 8 miles.

★ 10. While walking, a person can burn about 160 calories in an hour. Write and graph a function to show the number of calories a person burns while walking.

Test PREP

11. Which function is graphed at the right?

    a. $y = x + 3$
    b. $y = x - 3$
    c. $y = -x + 3$
    d. $y = -x - 3$

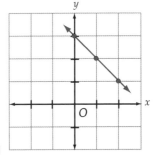

12. Which graph represents $y = -2x + 1$?

    a.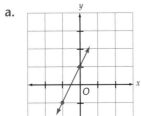
    b.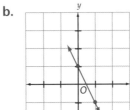

11.3 Graphing Linear Functions

11.4 Graphing Two Equations

Objective: to graph two equations on one coordinate plane

Kevin and Patsy Thomas are planning an ice-cream social. Kevin likes chocolate ice cream, and Patsy likes vanilla ice cream. Kevin buys 9 cartons of ice cream for the social. He buys twice as many cartons of vanilla as chocolate. How many packages of each does Kevin buy?

Let x represent the number of cartons of chocolate ice cream.

Let y represent the number of cartons of vanilla ice cream.

number of cartons of chocolate ice cream	plus	number of cartons of vanilla ice cream	equals	nine
x	+	y	=	9

number of cartons of vanilla ice cream	equals	twice	number of cartons of chocolate ice cream
y	=	2	x

Graph both equations on the same coordinate system. The lines intersect at (3, 6). Since this point lies on the graph of each equation, its coordinates are a solution to each equation. You can check this solution as shown below.

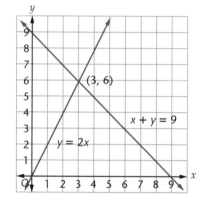

▶ Substitute 3 for x and 6 for y.

$x + y = 9$ $y = 2x$

$3 + 6 \stackrel{?}{=} 9$ $6 \stackrel{?}{=} 2 \cdot 3$

$9 = 9$ ✓ $6 = 6$ ✓

So, $x = 3$ and $y = 6$. Kevin buys 3 cartons of chocolate ice cream and 6 cartons of vanilla ice cream.

Try THESE

Copy and complete each table. Then graph the two equations in each problem on the same coordinate system. State the solution as an ordered pair.

1. $y = x - 1$ $y = 11 - x$

x	y
1	0
3	■
5	■

x	y
0	11
2	■
4	■

2. $y = x + 2$ $y = -x$

x	y
-3	■
-1	■
1	■

x	y
-1	■
0	■
1	■

Exercises

Graph the two equations in each exercise on the same coordinate system. State the solution as an ordered pair.

1. $y = 6 - x$
   $y = x - 2$
2. $y = 4x$
   $y = -x + 5$
3. $y = -x + 3$
   $y = 1 + x$
4. $y = 2x - 1$
   $y = x + 3$
5. $y = -x + 3$
   $y = -2x + 4$
6. $y = -2x + 7$
   $y = x - 5$

Graph the two equations in each exercise on the same coordinate system. Then state whether there is *one solution* or *no solutions*.

7. $y = 5$
   $y = 2$
8. $y = x - 10$
   $y = 11 - x$
9. $y = x + 1$
   $y = x + 4$

Problem SOLVING

★10. The perimeter of a rectangular garden is 46 meters. The length is 3 meters longer than the width. Write two equations to describe the situation. Graph the equations on the same coordinate system. What is the length and width of the garden?

★11. Graph the equation $y = 2x + 3$. At the point (2, 7), draw a line perpendicular to the first line. What is the equation of the perpendicular line?

Mid-Chapter REVIEW

Determine whether each sequence is *arithmetic, geometric,* or *neither*. If it is arithmetic or geometric, state the common difference or ratio.

1. 2, 6, 10, 14, ...
2. 3, 9, 27, 81, ...
3. 23, 17, 11, 5, ...

Find each function value.

4. $f(-5)$ if $f(x) = 3x + 6$
5. $f(3)$ if $f(x) = -x + 6$

Graph each function on graph paper.

6. $y = x - 4$
7. $y = 3x - 2$
8. $y = -x + 2$

Problem Solving

The Pyramid of Basketballs

Eddie and Raul are trying to make it into the *Around the World Book of Records*. They want to make the world's largest pyramid of basketballs. They have collected 10,000 basketballs. If each layer consists of a square number of basketballs, how many layers can they make? How many extra basketballs will they have?

Extension

Eddie and Raul can make one pyramid in which the total number of basketballs used is also a square number. How many layers does it have? (*Hint:* The answer is *not* one.)

Cumulative Review

Estimate.

1. $27.59 + 82.06$
2. $29{,}641 - 6{,}073$
3. 86.7×2.94
4. 0.406×0.029
5. $12.1\overline{)3.831}$

Compute.

6. $92.7 + 34.35$
7. $62.39 - 8.954$
8. 0.087×0.03
9. 0.968×3.49
10. $16\overline{)11}$

11. $15.5\overline{)96.1}$
12. $12 + \text{-}9$
13. $5 - 11.5$
14. $\text{-}25 - \text{-}15$
15. $\text{-}8.2 \bullet 9$
16. $14 \bullet \text{-}0.2$
17. $32 \div \text{-}8$
18. $\text{-}90 \div \text{-}30$

Compute. Write each answer in simplest form.

19. $\frac{5}{6} + \frac{5}{12}$
20. $4\frac{3}{4} + 6\frac{1}{8}$
21. $3\frac{3}{4} + 2\frac{1}{6}$
22. $\frac{11}{15} - \frac{2}{5}$
23. $7\frac{3}{4} - 4\frac{3}{8}$
24. $14\frac{2}{3} - 6\frac{3}{4}$
25. $\frac{7}{8} \bullet 2$
26. $4\frac{4}{9} \bullet \frac{3}{5}$
27. $2\frac{2}{7} \bullet 4\frac{1}{12}$
28. $\frac{7}{16} \div \frac{1}{4}$
29. $2\frac{5}{8} \div 1\frac{7}{8}$
30. $14 \div 5\frac{3}{5}$

Solve each equation. Check your solution.

31. $d + 40 = 27$
32. $a - 4 = 32$
33. $10z = 5$
34. $\frac{x}{7} = 12$
35. $\frac{2}{3}r = \text{-}24$
36. $\frac{y}{3} + 6 = \text{-}9$

Use the Pythagorean theorem to find the length of the hypotenuse of each right triangle. The lengths of the legs are given. Round your answers to the nearest tenth.

37. 6 mm, 8 mm
38. 9 ft, 48 ft
39. 8 m, 16 m
40. 12 in., 9 in.
41. 24 m, 7 m
42. 6 yd, 12 yd

Solve.

43. Find the perimeter and area of the rectangle shown at the right.

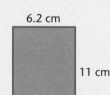
6.2 cm
11 cm

44. Find the surface area and volume of a rectangular prism whose base is 7 ft by 8 ft and whose height is 12 ft.

45. The price of a gallon of regular gasoline at five service stations was $1.299, $1.239, $1.339, $1.249, and $1.309. Find the average price to the nearest tenth of a cent.

46. Benito Diaz builds a toolbox. It is 1.1 m long, 0.6 m wide, and 0.4 m deep. Find its surface area and volume.

11.5 Slope of a Line

Objective: to find the slope of a line using the slope formula

At the Mountain Ski Resort, the slopes vary in degrees of steepness. The black diamond hills are the most challenging and have the steepest slope.

Slope is the ratio of the vertical distance, **rise**, over the horizontal distance, **run**. You can find the slope of a line by using the coordinates of any two points on the line. The points can be represented by the coordinates (x_1, y_1) and (x_2, y_2).

Slope of a Line

The slope of a line with coordinates (x_1, y_1) and (x_2, y_2) is

$$m = \frac{\text{change in } y}{\text{change in } x} = \frac{y_2 - y_1}{x_2 - x_1}.$$

Examples

A. Find the slope of the line that passes through $A(-1, 2)$ and $B(1, 3)$.

$m = \dfrac{y_2 - y_1}{x_2 - x_1}$ definition of slope

$(x_1, y_1) = (-1, 2)$

$m = \dfrac{3 - 2}{1 - -1} = \dfrac{1}{2}$ $(x_2, y_2) = (1, 3)$

The slope of the line is $\dfrac{1}{2}$.

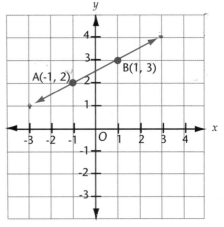

Lines can have a positive, negative, zero, or undefined slope.

Positive Slope — The line slants upward when you read the graph from left to right.

Negative Slope — The line slants downward when you read the graph from left to right.

Zero Slope — Every horizontal line has a slope of zero.

Undefined Slope — Every vertical line has an undefined slope.

B. Find the slope of the line that passes through A(2, 3) and B(3, 3).

$m = \dfrac{y_2 - y_1}{x_2 - x_1}$ definition of slope

$(x_1, y_1) = (2, 3)$

$m = \dfrac{3 - 3}{3 - 2} = \dfrac{0}{1} = 0$ $(x_2, y_2) = (3, 3)$

The slope of the line is 0.

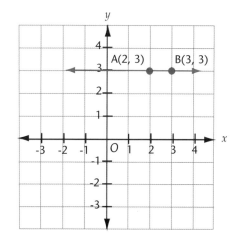

Try THESE

Find the slope of the line that passes through each pair of points.

1. (1, 5), (9, 9)
2. (-4, 3), (-1, 0)
3. (0, 3), (2, 7)

Exercises

Find the slope of the line that passes through each pair of points.

1. (3, -3), (1, -1)
2. (4, 4), (6, 8)
3. (-1, -4), (-3, 0)
4. (-1, 5), (-1, 3)
5. (7, 4), (2, 0)
6. (-2, 7), (8, 5)
7. (3, -8), (4, -6)
8. (-3, 7), (0, 5)
9. (-2, -2), (3, 1)
10. (-3, -4), (2, -4)
11. (-1, -3), (1, 2)
12. (-7, 4), (5, -3)

Problem SOLVING

★13. Lines *a* and *b* are parallel. Find the slope of each line. What do you notice about the slopes of lines *a* and *b*? Can you make a conjecture about the slopes of parallel lines?

★14. Lines *c* and *d* are perpendicular. Find the slope of each line. What do you notice about the slopes of lines *c* and *d*? Can you make a conjecture about the slopes of perpendicular lines?

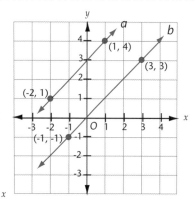

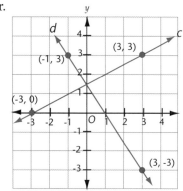

11.5 Slope of a Line

11.6 Slope-Intercept Form

Objective: to graph linear equations using the slope and the y-intercept

You can find the slope of a line by looking at the graph of the line. You can also find the slope of a line by looking at the equation of the line.

An equation written in the form of $y = mx + b$ is written in **slope-intercept form**. When an equation is written in slope-intercept form, m is the slope and b is the y-intercept.

slope ⟶ $y = mx + b$ ⟵ y-intercept

Examples

State the slope and the y-intercept of each equation.

A. $y = -3x - 5$

$y = -3x + -5$ Rewrite $-3x - 5$ as $-3x + -5$.

$y = mx + b$ $m = -3$ and $b = -5$

The slope is -3, and the y-intercept is -5.

B. $2x + y = 4$

$2x - 2x + y = -2x + 4$

$y = -2x + 4$ Solve the equation for y.

$y = mx + b$ $m = -2$ and $b = 4$

The slope is -2, and the y-intercept is 4.

C. Graph $y = \frac{1}{2}x - 2$ using the slope and the y-intercept.

- Step 1: Find the slope. $y = \frac{1}{2}x - 2$

  $m = \frac{1}{2}$ and $b = -2$

- Step 2: Graph the y-intercept. (0, -2)

- Step 3: Start at the y-intercept. Move according to the slope.

  1 change in y—move up 1 unit

  2 changes in x—move right 2 units

- Step 4: Connect the points with a line.

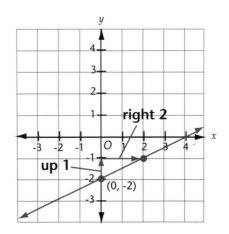

Try THESE

State the slope and the y-intercept for the graph of each equation.

1. $y = -5x + 4$
2. $y = \frac{3}{4}x + 6$
3. $y = 4x - 7$

Exercises

State the slope and the y-intercept for the graph of each equation.

1. $y = x + 6$
2. $y = \frac{1}{2}x - 5$
3. $y = \frac{2}{3}x + 4$

4. $y = -4x + 5$
5. $y = \frac{3}{4}x - 2$
6. $y = -x - 10$

7. $3x + y = 5$
8. $y - 5x = 9$
9. $2y - 4x = 8$

Graph each equation using the slope and y-intercept.

10. $y = x + 5$
11. $y = -2x + 3$
12. $y = -x + 3$

13. $y = \frac{1}{2}x + 2$
14. $y = 3x - 5$
15. $y = -\frac{3}{4}x + 5$

16. $y = -4x + 6$
17. $y - 2x = -1$
18. $3x + y = 4$

19. Write an equation in slope-intercept form with a slope of 3 and a y-intercept of -2.

20. Write an equation in slope-intercept form with a slope of -1 and a y-intercept of 5.

Problem SOLVING

21. What is the slope of a horizontal line? What is the y-intercept of a horizontal line?

Constructed RESPONSE

22. Luisa has a job at a shoe store. Luisa's pay for 1 day can be represented by the equation $y = 0.30x + 20$, where x is the total amount of Luisa's sales for a day.

   a. What does the slope represent?
   b. What does the y-intercept represent?

11.7 Graphing Inequalities

Objective: to graph inequalities on a coordinate plane

The graph of the linear equation $y = x + 1$ separates the coordinate plane into two regions.

The graph of the inequality $y > x + 1$ is the region *above* and *not including* the line.

Test a point to see which region satisfies the inequality. Try (0, 0).

$y > x + 1$
$0 \overset{?}{>} 0 + 1$
$0 > 1$ no

The side that does *not* contain (0, 0) is shaded.

The graph of the inequality $y \leq x + 1$ is the region *below* and *including* the line.

Test a point to see which region satisfies the inequality. Try (0, 0).

$y \leq x + 1$
$0 \overset{?}{\leq} 0 + 1$
$0 \leq 1$ yes

The side that contains (0, 0) is shaded.

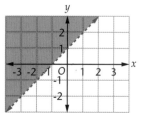
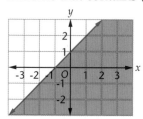

Graph $y > 2x - 3$.

Step 1	Step 2	Step 3
Find several ordered pairs that satisfy $y > 2x - 3$. | x	y | |---|---| | -1 | -5 | | 0 | -3 | | 1 | -1 | | 2 | 1 |	Graph the equation. Draw a dashed line because the line is not part of the graph.	Test (1, 1). Shade the region that includes (1, 1).

Try THESE

Graph each inequality.

1. $y < 2x$
2. $y > x - 1$
3. $y \leq x + 5$
4. $y \geq -2x$

Exercises

Graph each inequality.

1. $y \leq x + 3$
2. $y > -3x$
3. $y < x + 4$
4. $y < x - 5$
5. $y \geq 2x + 3$
6. $y > \frac{1}{2}x + 2$
7. $y > -2x - 2$
8. $y \geq 4x - 1$
9. $y \leq -x + 3$
10. $y < \frac{3}{4}x + 1$
11. $y \leq \frac{2}{3}x - 3$
12. $y < 5x - 1$
13. $y \geq -\frac{1}{2}x + 3$
14. $x + y < 2$ ★
15. $x + y > 2$ ★

Problem SOLVING

16. Graph the inequality "The sum of two numbers is less than three."
17. Graph the inequality "The difference of two numbers is greater than or equal to nine."
18. Laura has 45 minutes to run and swim.
    a. Make a graph showing all the amounts of time Laura can spend on running and swimming.
    b. Give two possible ways Laura can split her time between running and swimming.

Mind BUILDER

Second-Degree Equations

You can find ordered pairs that make second-degree equations true.
Consider the equation $y = x^2 - 1$.

$y = x^2 - 1$ Suppose $x = 2$.
$y = 2 \cdot 2 - 1$
$y = 4 - 1$
$y = 3$ The ordered pair is (2, 3).

Copy and complete each table.

1. $y = x^2 + 2$

x	y
-1	3
0	2
1	■
2	■

2. $y = 1 - x^2$

x	y
-1	0
0	1
1	■
2	■

3. $y = 2x^2$

x	y
-1	2
0	■
1	■
2	■

4. $y = 3x^2 - 4$

x	y
-1	■
0	■
1	■
2	■

11.8 Problem-Solving Strategy: Using Graphs

Objective: to solve problems using graphs

A small plane can carry 2,060 pounds of people, cargo, and fuel. The more people and cargo it carries, the less fuel it can carry. Reducing the amount of fuel decreases the flying time. A graph of the pounds of fuel needed to fly a certain number of hours is shown at the right.

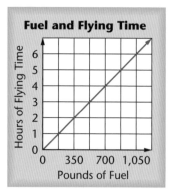

Suppose a plane carries 410 pounds of cargo and six people who weigh a total of 950 pounds. The rest of the weight is fuel. How many hours of flying time does the plane have?

You know the total weight the plane can carry (2,060 pounds) and the weight of the people and cargo. Find the weight of the fuel and how much flying time the amount of fuel allows.

First find how many pounds of fuel the plane can carry. Then use the graph to find the hours of flying time.

```
  410  pounds of cargo         2,060  total pounds allowed
+ 950  pounds of people      − 1,360  pounds of cargo and people
1,360                            700  pounds of fuel
```

Find 700 on the horizontal scale of the graph. Then move up to the line. From this point, move left and read the hours of flying time. The plane has 4 hours of flying time.

You can check the solution by working backwards. Assume that the plane has 4 hours of flying time. Find how much fuel is needed and the pounds of people and cargo allowed. The solution is correct.

Try THESE

Solve. Use the information and the graph above.

1. A plane carries 770 pounds of cargo and 4 people who weigh a total of 765 pounds. The rest of the weight is fuel. How many hours of flying time does the plane have?

2. A plane must fly 4 hours to reach its destination. The pilot and copilot each weigh 145 pounds. There are no passengers. How much cargo can the plane carry?

Solve

Use the graph to solve.

Mr. Lewis earns a base salary of $300 per week plus a commission of 5% on all sales.

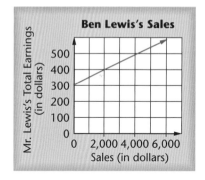

1. What are Mr. Lewis's total earnings if his sales are $1,000?

2. Mr. Lewis's goal is to make $450 this week. How much must his sales be?

★ 3. Write an equation in the form of $y = mx + b$ to describe the graph. Use x for Mr. Lewis's sales and y for his total earnings.

4. Why doesn't the graph begin at $x = 0$?

Use the chart to solve.

Ace Repair Shop has a chart for its employees to use in computing customers' bills.

Hours	0.5	0.75	1	1.25	1.5	2
Charge	$29.00	$33.50	$38.00	$42.50	$47.00	$56.00

5. Draw a graph to show the relationship given in the chart.

6. Ace charges a flat fee plus a per hour fee. What is the flat fee?

7. What is the fee per hour?

8. What is the charge for a job that takes 2.25 hours?

Use any method to solve.

9. Cranberries can be picked by mechanical harvester machines that can pick 10,000 pounds of cranberries each day. How many tons of cranberries can be harvested by 3 machines in 5 days?

★ 10. Nely said, "I am thinking of a number that is both a perfect square and a perfect cube. The numbers 1 and 729 satisfy both conditions, but this number is between those two." What number is Nely thinking of?

11. The perimeter of a certain tennis court is 76 yards. The length of the tennis court is 2 yards more than twice the width. What is the length of the tennis court?

12. Michelle's birthday was on a Monday in 2001. On what day of the week was her birthday in 2003?

★ 13. The Martinez family is moving from St. Louis, Missouri, to Houston, Texas, a distance of 779 miles. Their car averages 32 miles per gallon of gasoline. The moving van averages 5 miles per gallon of gasoline. How many more gallons of gasoline will the van use than the car as they make the trip together?

11.8 Problem-Solving Strategy: Using Graphs

Chapter 11 Review

Language and Concepts

Choose the correct term to complete each sentence.

1. The horizontal number line in a coordinate system is called the _____.
2. A(n) _____ is a list of numbers that follows a pattern.
3. The coordinates of points are named as _____.
4. In a(n) _____, there is exactly one output value for every input value.
5. A rule that describes the relationship between two sets of numbers is called a(n) _____.
6. The _____ of a relation is the set of all second components in a set of ordered pairs.
7. _____ is a ratio of vertical distance to horizontal distance.
8. The equation $y = 2x + 1$ is written in _____.
9. A(n) _____ is a sequence where the difference between any two consecutive terms is the same

ordered pairs
range
relation
slope
slope-intercept form
x-axis
sequence
arithmetic sequence
geometric sequence
function
domain
y-axis

Skills and Problem Solving

Determine whether each sequence is *arithmetic, geometric,* or *neither.* If it is arithmetic or geometric, state the common difference or ratio. Find the next three terms in each sequence. Section 11.1

10. 15, 45, 135, 405, . . .
11. 2, 4, 8, 32, . . .
12. 6, 2, -2, -6, . . .

Find the value of each function. Section 11.2

13. $f(7)$ if $f(x) = 2x - 6$
14. $f(-5)$ if $f(x) = 8 + 3x$
15. $f(\frac{1}{2})$ if $f(x) = 9x - 2$

Graph each function on graph paper. Section 11.3

16. $y = 3 + x$
17. $y = x - 1$
18. $y = 2x + 1$

Graph the two equations in each problem on the same coordinate system. State the solution as an ordered pair. Section 11.4

19. $y = 2x - 1$
    $y = x + 4$

20. $y = 3x$
    $y = x$

21. $y = 3x - 2$
    $x + y = 6$

Find the slope of the line that passes through each pair of points. Section 11.5

22. (-1, -3), (-6, -7)
23. (5, 3), (5, -4)
24. (6, 5), (-2, 1)

State the slope and the *y*-intercept for the graph of each equation. Then graph each equation using the slope and *y*-intercept. Section 11.6

25. $y = -5x - 7$
26. $y = \dfrac{3}{4}x + 2$
27. $3x + y = 7$

Graph each inequality. Section 11.7

28. $y < x + 2$
29. $y \geq \dfrac{1}{2}x - 3$
30. $y < -2x + 4$

Solve. Section 11.8

31. Each game of bowling at Al's Lanes costs $1.25. Draw a graph to show this relationship.

32. Neva can spend $4.00 on bowling at Al's Lanes. She rents shoes for 25¢. How many games can she bowl?

Chapter 11 Review

Chapter 11 Test

Determine whether each sequence is *arithmetic, geometric,* or *neither*. If it is arithmetic or geometric, state the common difference or ratio. Find the next three terms in each sequence.

1. 5, 2, -1, -4, ...
2. 25, 36, 49, 64, ...
3. 32, -8, 2, $-\frac{1}{2}$, ...

Find the value of each function.

4. $f(-4)$ if $f(x) = -3x - 5$
5. $f(3)$ if $f(x) = \frac{x}{3} + 2$

Graph each function or inequality on graph paper.

6. $y = -3x + 5$
7. $y = 3 - x$
8. $y > \frac{3}{4}x + 2$

9. Graph $y = x + 2$ and $y = 3x$ on the same coordinate system. State the solution as an ordered pair.

Find the slope of the line that passes through each pair of points.

10. (-3, 4), (-2, 6)
11. (1, -4), (2, -4)
12. (5, 3), (7, 1)

Solve.

13. The cost for dance lessons is $12 an hour plus a registration fee of $20. The cost y for x hours is $y = 12x + 20$.

    a. Graph this equation.
    b. State the *y*-intercept of the equation. What does it represent?
    c. State the slope of the equation. What does it represent?

14. The cost of telephone service in Branson is related to the number of calls a customer makes.

Calls	0	5	10	20	40
Cost	$12	$12.50	$13	$14	$16

    a. Draw a graph to show this relationship.
    b. Mr. Pantera's bill is $13.80. Use the graph to find how many calls were made on his telephone.

Change of Pace

Rule of 78

The interest on a loan is frequently based on the unpaid balance (the amount of the loan minus the amount already paid). More interest is owed the first month than the last month of the loan.

Suppose a 1-year loan has a total interest charge of $72.00. For a 1-year loan, the following sum is used to figure the amount of interest each month.

$12 + 11 + 10 + 9 + 8 + 7 + 6 + 5 + 4 + 3 + 2 + 1 = 78$

You can find the amount of interest owed as follows:

after 1 month: $\frac{12}{78}$ of interest owed ▶ $\frac{12}{78} \times \$72 = \$11.08$

after 3 months: $\frac{12}{78} + \frac{11}{78} + \frac{10}{78}$, or $\frac{33}{78}$, of interest owed ▶ $\frac{33}{78} \cdot \$72 = \$30.46$

after 6 months: $\frac{12}{78} + \frac{11}{78} + \frac{10}{78} + \frac{9}{78} + \frac{8}{78} + \frac{7}{78}$, or $\frac{57}{78}$ ▶ $\frac{57}{78} \cdot \$72 = \$52.62$

The above method is called the **rule of 78**. Do you see why?

Use the rule of 78 to find the part of the interest owed on a 1-year loan for each of the following. The total amount of interest is $72.00. Round amounts to the nearest cent.

1. first 2 months
2. the ninth month
3. first 8 months
4. last 6 months

Solve. Round amounts to the nearest cent.

5. The total amount of interest on a 1-year loan is $96.00. Find the interest owed for the first month.

6. The amount of interest owed on a 1-year loan is $195.00. Find the interest owed for the first 4 months.

7. Gail Knisely owes $132.00 in interest for the first 3 months of a 1-year loan. Find the total amount of interest for the entire 1-year loan.

8. Jud Harris owes $12.00 in interest for the last 3 months of a 1-year loan. Find the total amount of interest for the entire 1-year loan.

Cumulative Test

1. If $y = 2x - 4$, what number completes the table?

x	y
14	24
20	36
26	■

    a. 40
    b. 44
    c. 48
    d. 52

2. By what are the triangles congruent?

    a. ASA
    b. SAS
    c. SSS
    d. none of the above

3. What is the volume of the cylinder?

    a. about 502.4 in.^{3}
    b. about 602.88 in.^{3}
    c. about 1,004.8 in.^{3}
    d. about 4,019.8 in.^{3}

4. What is the formula for the area (A) of a square with sides s units long?

    a. $A = s \times s$
    b. $A = s + s$
    c. $A = 2 \times s$
    d. $A = 4 \times s$

5. Find the circumference of this circle. Use $\pi \approx \frac{22}{7}$.

    a. 110 cm
    b. 220 cm
    c. 770 cm
    d. 3,850 cm

6. If $y = -2x + 1$, what number completes the table?

x	y
-2	5
1	-1
3	■

    a. -7
    b. -5
    c. -3
    d. 4

7. Dora spends 2¢ on the first day and 4¢ on the second day. Each day thereafter she spends twice as much as the previous day. How much will she spend on the tenth day? Which of the following tells how to solve the problem?

    a. $2 + 4 + 8 + 16 + \ldots$
    b. $2 + 4 + 6 + 8 + \ldots$
    c. 2^9
    d. 2^{10}

8. Which point could have the coordinates (-2, 2)?

    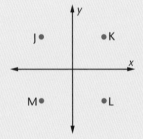

    a. J
    b. L
    c. K
    d. M

9. In the figure at the right, what is the area of the unshaded portion?

    a. 60 ft^2
    b. 88 ft^2
    c. 260 ft^2
    d. 320 ft^2

10. Suppose you multiply the x-coordinates of each point by -1. Which will the resulting figure be?

    a. reflection
    b. translation
    c. rotation
    d. none of the above

    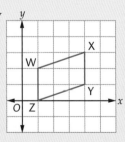

CHAPTER 12

Algebraic Relationships

Catherine Beehler
Calvert Day School

12.1 Nonlinear Functions

Objective: to determine whether a function is linear or nonlinear

The graph of a linear function is a straight line. A linear function has a constant rate of change. When you graph a **nonlinear function**, the graph is not a straight line because it does not have a constant rate of change.

You can tell if a function is linear or nonlinear by looking at the graph of the function, the equation of the function, or the function table.

Examples

Determine whether each graph represents a *linear function* or a *nonlinear function*.

A.

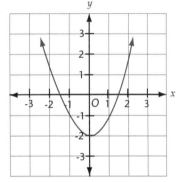

The graph is not a straight line, therefore it is nonlinear. The graph is a curve.

B.

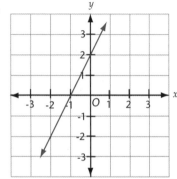

The graph is a straight line. It is linear.

You can tell by looking at the equation of a function whether it is linear or not. You know that a linear function is in the form of $y = mx + b$.

Determine whether each equation represents a *linear function* or a *nonlinear function*.

C. $y = x + 7$

The equation represents a linear function. It is in the form of $y = mx + b$.

D. $y = 3x^2 - 5$

The equation does not represent a linear function. It cannot be written in the form of $y = mx + b$.

A linear function has a constant rate of change. A nonlinear function does not.

Determine whether each table represents a *linear function* or a *nonlinear function*.

E.

x	y
1	1
2	4
3	9
4	16

This is a nonlinear function.

The rate of change is not constant.

As x increases by 1, y increases by more each time.

F.

x	y
1	3
2	6
3	9
4	12

This is a linear function.

The rate of change is constant.

As x increases by 1, y increases by 3 each time.

Try THESE

Determine whether each graph represents a *linear function* or a *nonlinear function*. Explain.

1.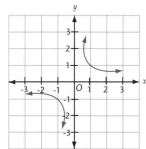

2.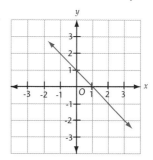

Exercises

Determine whether each equation represents a *linear function* or a *nonlinear function*. Explain.

1. $y = 6x - 8$
2. $y = \frac{5}{x}$
3. $y = 3$
4. $y = 3x^3 + 1$
5. $y = 2x$
6. $y - x = 6$

Determine whether each table represents a *linear function* or a *nonlinear function*. Explain.

7.
x	y
2	8
5	20
8	32
11	44

8.
x	y
-1	-2
0	2
1	4
2	6

9.
x	y
-3	9
-1	3
1	-3
3	-9

Problem SOLVING

10. Is the formula for finding the perimeter of a square linear or nonlinear?

11. Is the formula for finding the area of a square linear or nonlinear?

12.2 Quadratic Functions

Objective: to graph nonlinear functions

Keith kicks a soccer ball and watches it fly up in the air and then fall down. The path the soccer ball takes can be graphed.

The graph of the soccer ball function is not linear. It does not form a straight line. It is a **quadratic function**. A quadratic function is a function in which the greatest power of the variable is 2. The graph of a quadratic function is a curve that opens upward or downward. The curve is called a **parabola**.

You can graph a quadratic function by making a table and plotting the points.

Example

Graph the equation $y = x^2$.

Step 1	Step 2	Step 3
Find several ordered pairs that satisfy the equation.	Graph those ordered pairs.	Connect the points with a smooth curve.

Step 1:

x	y	(x, y)
-3	9	(-3, 9)
-2	4	(-2, 4)
-1	1	(-1, 1)
0	0	(0, 0)
1	1	(1, 1)
2	4	(2, 4)
3	9	(3, 9)

308   12.2 Quadratic Functions

Try THESE

Copy and complete each table. Graph each function on graph paper.

1. $y = x^2 + 2$

x	y
-1	3
0	2
1	■
2	■

2. $y = 1 - x^2$

x	y
-1	0
0	1
1	■
2	■

3. $y = 2x^2$

x	y
-1	2
0	■
1	■
2	■

Exercises

Graph each function on graph paper.

1. $y = x^2 + 1$
2. $y = x^2 - 3$
3. $y = 5x^2$
4. $y = 2x^2 - 1$
5. $y = 3x^2 - 4$
6. $y = 3 - x^2$
7. $y = -3x^2$
8. $y = x^2 - 5$
9. $y = 2.5x^2$
10. $y = -2x^2 + 4$
★11. $y = \frac{1}{2}x^2 - 2$
★12. $y = -x^2 + 4$

Problem SOLVING

13. The function $d = \frac{1}{2}at^2$ represents the distance d that an object travels in time t with a given acceleration rate a. A rocket accelerates at a rate of 30 feet per second every second.
    a. Graph $d = \frac{1}{2}(30t^2)$.
    b. Find the distance traveled by the rocket after 5 seconds.
    c. About how long will it take the rocket to travel 60 feet?

Constructed RESPONSE

14. The function $d = -16t^2 + h$ models the distance d from the ground of an object falling at time t, in seconds, from height h. In science class, you drop an egg into a carton from a height of 150 feet.
    a. Graph $d = -16t^2 + 150$.
    b. Find the height of the egg after 1 second. Find the height of the egg after 3 seconds.
    c. When will the egg hit the carton on the ground? Explain.

12.3 Polynomials

Objective: to simplify polynomials

To simplify an expression with like terms, you combine the like terms by adding or subtracting. The terms in an expression are called **monomials**. An algebraic expression with one or more monomials linked by addition or subtraction is called a **polynomial**.

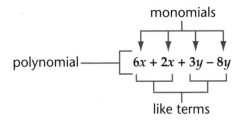

Examples

A. Simplify. $5a + 3b - b + 4a$

$5a + 3b + -1b + 4a =$	Change subtraction to adding a negative.
$(5a + 4a) + (3b + -1b) =$	Group the like terms.
$9a + 2b$	Simplify by combining the like terms.

B. Simplify. $-3x^2 + 5x + 7x^2$

$-3x^2 + 5x + 7x^2 =$	
$(-3x^2 + 7x^2) + 5x =$	Group the like terms.
$4x^2 + 5x$	Simplify by combining the like terms.

Like terms must have the same variables and powers. $4x^2$ and $4x$ are not like terms because x^2 and x do not have the same power.

C. Simplify. $3x + 2x^2 - 5 + x + 3$

$3x + 2x^2 + -5 + 1x + 3 =$	Change subtraction to adding a negative.
$2x^2 + (3x + 1x) + (-5 + 3) =$	Group the like terms.
$2x^2 + 4x + -2$ or $2x^2 + 4x - 2$	Simplify.

Try THESE

Simplify each polynomial.

1. $2x + 5 + 9$

2. $-3x^2 + 8 + 5x^2$

3. $7 - 5a + a + 2$

Exercises

Simplify each polynomial. If the polynomial cannot be simplified, write *simplest form*.

1. $6g + 8k - 9g + 2k$
2. $-2s + 8t - 10s - 5t$
3. $b + a - 5a + 3b$
4. $7x + 8y - 10x - 9y$
5. $3x^2 + 8x + 7x$
6. $-4x + 6x^2 + 5x$
7. $9a^2 + 6b - 5a$
8. $4r + 8t + 7r - 9t$
9. $-6x^2 + 5x + x^2$
10. $a^2 + 5a - 3a^2$
11. $4 - 3x^2 + 5x + 8$
12. $3w^2 + 5w - 8w + 3$
13. $b^2 + 8b - 7b + b$
14. $7s + 3s^2 + 2s - 8s^2$
15. $-5x^2 + 6x - 9x^2 + 2$
16. $6w + 4w^2 - 8w + 3w - 9w^2$
17. $a^2 + 7a^2 + 6a + 7a - 5 + 3$
18. $-4d + 7d + 8d^2 + 9d^2 - 6$
19. $-5w^3 + 6w^2 + 7w^2 - 6w + 4w^3$
20. $10 + 3a^2 - 8a^3 + 7a^3 - 9 - 2a^2$
21. $-3b^2 + 8b + 6b + 7b^3 - 9b^2 + 4b^2$
22. $3.5w + 2.1w^2 + 9.8 + 5.6w + 1.2$
23. $3\frac{1}{2}x + \frac{3}{4}x^2 - 2x + \frac{1}{4}x^2$

Problem SOLVING

24. Jane has 3 gallons of apple juice and 2 gallons of orange juice. She buys 2 more gallons of apple juice and 5 more gallons of orange juice. Write an expression with four terms that represents the number of gallons of apple juice *a* and orange juice *o* she has. Then simplify the expression.

25. For what values of x does $3x^2 + 6x = 9x^2$? Why does it not work for all values of x?

Test PREP

26. Simplify. $2x^2 + 6x - 7x + 8 - 3x^2 + 1$
    a. $2x^2 + x + 9$
    b. $-2x^2 - x - 9$
    c. $-x^2 - x + 9$
    d. $x^2 + 9$

27. Simplify. $-a^3 + 4a^2 - 7a^2 - 5a^3$
    a. $-6a^3 - 3a^2$
    b. $-8a^3 - a^2$
    c. $3a^2 - 12a^3$
    d. $6a^3 + 3a^2$

Mid-Chapter REVIEW

Determine whether each equation represents a *linear function* or a *nonlinear function*.

1. $y = 3x + 5$
2. $y = x^2 - 4$
3. $y = \frac{1}{2}x$

Graph each function on graph paper.

4. $y = x^2 + 3$
5. $y = 2x^2 - 4$
6. $y = \frac{1}{2}x^2 + 1$

Problem Solving

Undo the Magic

The magic sum is 0. Make a 3-by-3 square. Place integers in all squares so that each row, column, and diagonal has the magic sum. Each integer must be used only once. The integers used must be consecutive (for example, from -10 to -2).

Next try a magic sum of 3.

Extension

Use consecutive whole numbers from 1 through 9 to complete the magic square at the right. Or, explain why it is impossible to complete.

Cumulative Review

Estimate.

1. 4.96 + 7.854
2. 65.85 − 19.09
3. 52.7 × 6.9
4. 398)2,145
5. 9.2)284.6

Compute.

6. $\frac{1}{8} + \frac{1}{2}$
7. $2\frac{4}{5} + \frac{1}{2}$
8. $-5 - 2\frac{1}{5}$
9. $7\frac{1}{9} - 4\frac{5}{6}$
10. $6 \cdot \frac{3}{4}$
11. $2\frac{1}{3} \cdot \frac{2}{5}$
12. $2\frac{3}{5} \div \frac{2}{5}$
13. $2\frac{2}{3} \div 2\frac{5}{6}$
14. $2 - -12$
15. $-41 + 3$
16. $-6.4 + 8.13$
17. $-9.36 - 53.2$
18. $8 \cdot -4$
19. $-0.4 \cdot -1.7$
20. $56 \div -7$
21. $-9.5 \div 0.19$
22. $-2\frac{3}{4} + 4\frac{5}{6}$
23. $\frac{4}{9} - \frac{1}{3}$
24. $-3\frac{1}{2} \cdot 3\frac{1}{3}$
25. $9 \div -\frac{3}{8}$

Find the greatest common factor (GCF) of each group of numbers.

26. 14, 21
27. 24, 36
28. 6, 24, 39

Find the least common multiple (LCM) of each group of numbers.

29. 10, 25
30. 4, 16
31. 6, 8, 10

Complete. Use the congruent triangles shown.

32. $\overline{YZ} \cong$ ■
33. $\angle PMN \cong$ ■
34. $\angle XZY \cong$ ■
35. $\overline{MP} \cong$ ■

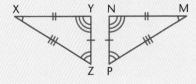

Replace each ● with >, <, or = to make a true statement.

36. 5.7 ● 5.76
37. −14 ● 1
38. −2.3 ● −2.30
39. $|-5| - |-2|$ ● $|5 - 2|$
40. 0.30 ● $\frac{1}{3}$
41. $\frac{3}{4}$ ● 0.755

Solve.

42. Ruth's car can travel 23 miles per gallon. It has a 27-gallon fuel tank. How many miles can it travel on 1 tank of gasoline?
43. One day the temperature reading changed from 11°C to a reading 14° lower. What was the new reading?
44. What number is 65% of 60?
45. What percent of 75 is 30?
46. 70 is 35% of what number?
47. 5.4% of 80 is what number?

Find the range, mean, median, and mode of each set of numbers.

48. 36, 28, 35, 37, 24, 35, 29
49. 384, 638, 596, 972, 430

12.4 Adding and Subtracting Polynomials

Objective: to add and subtract polynomials

You can add and subtract polynomials in two different ways. You can add and subtract polynomials vertically or horizontally by combining like terms. Both methods give the same answer.

Examples

A. Add. $(5x + 6) + (2x - 7)$

Vertical Method

$5x + 6$ Line up the like terms.
$+ 2x - 7$
$7x - 1$ Then add.

Horizontal Method

$(5x + 6) + (2x - 7)$ Use the commutative and associative properties of addition to group the like terms.
$= (5x + 2x) + (6 - 7)$
$7x \quad - \quad 1$

The sum is $7x - 1$.

B. Subtract. $(8a^2 + 6a + 3) - (5a^2 - 3a + 7)$

You can subtract the two polynomials, or you can add the inverse of the second polynomial to the first polynomial.

Vertical Method

$8a^2 + 6a + 3$ Line up the like terms.
$- \ 5a^2 - 3a + 7$
$3a^2 + 9a - 4$ Then subtract.

or

$8a^2 + 6a + 3$ To subtract $5a^2 - 3a + 7$,
$+ \ -5a^2 + 3a - 7$ add $-5a^2 + 3a - 7$.
$3a^2 + 9a - 4$

Horizontal Method

$(8a^2 + 6a + 3) - (5a^2 - 3a + 7)$
$= (8a^2 + 6a + 3) + (-5a^2 + 3a + -7)$ To subtract, add the opposite.
$= (8a^2 + -5a^2) + (6a + 3a) + (3 + -7)$ Group the like terms.
$= 3a^2 + 9a - 4$ Then add.

The difference is $3a^2 + 9a - 4$.

Try THESE

Add or subtract.

1. $8x + 7$
   $- \ 3x + 4$

2. $4x^2 + 9x - 5$
   $+ \ 2x^2 - 8x + 4$

3. $7r^2 + 3r - 1$
   $+ \ 2r^2 + 5$

Exercises

Add.

1. $(9m + 6) + (-5m - 6)$
2. $(-5w - 4) + (-3w - 10)$
3. $(-4p + 10) + (3p - 7)$
4. $(3r^2 + 7r + 1) + (4r^2 - 8r - 2)$
5. $(5e^2 + 8e - 7) + (-7e^2 + e + 2)$
6. $(8w^2 + 9w + 15) + (4w - 8)$
7. $(-3p^2 - 5p + 7) + (-7p^2 + 10p - 3)$
8. $(k^2 + 7) + (4k^2 + 5k)$

Subtract.

9. $(2c - 4) - (3c + 2)$
10. $(5p^2 + 3) - (8p^2 + 1)$
11. $(3c^2 + 7) - (-8c^2 + 1)$
12. $(6h + 1) - (9h + 4)$
13. $(5r^2 + 7r - 8) - (5r^2 + 3r - 4)$
14. $(c^2 + 4c - 7) - (3c^2 - 5c - 9)$
15. $(-7w^2 - 2w - 1) - (-5w^2 + 3w - 2)$
16. $(4a^2 + 8a + 2) - (a^2 + 3a)$

Add or subtract. Then evaluate for $a = 4$ and $b = -2$.

17. $(3a + 2) + (5a - 5)$
18. $(-4b - 7) - (2a + 3)$
19. $(2a + 5b) + (7a + 8b)$
20. $(-3a + 6b) - (-4a - 8b)$

Problem SOLVING

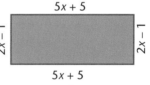

21. Look at the rectangle to the right. Write a polynomial expression in simplest form for the perimeter of the rectangle.

★22. Look at the triangles to the right. Write an expression in simplest form that represents the difference between the perimeter of the larger triangle and the perimeter of the smaller triangle.

Mind BUILDER

Magic Squares

The rows, columns, and diagonals of magic squares all have the same sum. The magic sum for this square is 21.

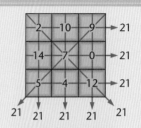

Find the value of each variable in each magic square. What is the magic sum?

1.

2.

a	b	1.8	5.8
c	5	4.6	3.8
d	3.4	3	e
2.2	6.2	f	g

3.

h	12	u	4	w
k	m	n	7	8
2	3	p	10	11
5	6	q	$10\frac{1}{2}$	r
$5\frac{1}{2}$	9	s	1	$4\frac{1}{2}$

4. Make your own magic square. What is the magic sum?

12.5 Multiplying Polynomials

Objective: to multiply polynomials

A polynomial with only one term is called a monomial. You can multiply monomials. To multiply monomials, you multiply the coefficients and add the exponents of the powers with the same bases.

Example
Multiply. $(9x^2)(-4x^6)$

$9x^2 \cdot -4x^6 =$	9 is the coefficient of x^2, and -4 is the coefficient of x^6.
$(9 \cdot -4)(x^2 \cdot x^6) =$	Use the commutative and associative properties.
$(-36)(x^{2+6}) =$	Multiply the coefficients. Add the exponents.
$-36x^8$	Simplify.

A polynomial with two terms is called a binomial. You can multiply a monomial and a binomial. You can use the distributive property to multiply a monomial and a binomial.

$$8(x + 3) = 8 \cdot x + 8 \cdot 3$$ — Distribute the 8 to each term in the parentheses.
$$= 8x + 24$$ — Simplify.

(binomial: $x+3$; monomial: 8)

More Examples

A. Multiply. $y(3y + 7)$

$y(3y + 7) =$
$y \cdot 3y + y \cdot 7 =$ Use the distributive property.
$3y^2 + 7y$ Simplify.

B. Multiply. $3n(n + 6)$

$3n(n + 6) =$
$3n \cdot n + 3n \cdot 6 =$ Use the distributive property.
$3n^2 + 18n$ Simplify.

C. Multiply. $-4w(3w^2 - 5)$

$-4w(3w^2 - 5) = -4w(3w^2 + -5) =$ Rewrite $3w^2 - 5$ as $3w^2 + -5$.
$-4w \cdot 3w^2 + -4w \cdot -5 =$ Use the distributive property.
$-12w^3 + 20w$ $w \cdot w^2 = w^{1+2} = w^3$

D. Multiply. $5x(x^2 + 3x + 6)$

$5x(x^2 + 3x + 6) =$
$5x \cdot x^2 + 5x \cdot 3x + 5x \cdot 6 =$ Use the distributive property.
$5x^3 + 15x^2 + 30x$ Simplify.

Try THESE

Multiply.

1. $(2x)(10x)$
2. $(-3x^2y)(-5y^2)$
3. $x(x + 6)$

Exercises

Multiply.

1. $5x(x + 2)$
2. $-3m(m + 8)$
3. $4d(2d - 9)$
4. $y(3y^2 + 8)$
5. $-2r(8r + 5)$
6. $-8w(6w^2 - 7)$
7. $x(x^2 + 5x + 7)$
8. $m(2m^2 - 3m - 6)$
9. $4a(a^2 + 7a + 4)$
10. $8w(3w^2 + 7w - 2)$
11. $-4x(x^2 - 5x - 4)$
12. $-7m(5m^2 + 3m - 1)$

Problem SOLVING

13. Write an expression in simplest form for the area of the parallelogram to the right. Find the area of the parallelogram if $x = 6$.

14. Write an expression in simplest form for the area of the triangle to the right. Find the area of the triangle if $x = 9$.

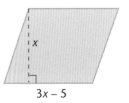

Constructed RESPONSE

15. Monica and Phoebe are finding the product of $4x$ and $3x^2 - 8x + 5$. Who is correct? Explain.

Monica
$4x(3x^2 - 8x + 5) = 12x^3 - 32x^2 + 20x$

Phoebe
$4x(3x^2 - 8x + 5) = 12x^2 - 32x + 20$

MiXeD REVIEW

Compute.

16. $6^2 + 3 \cdot -5$
17. $4 \cdot 2 + (-8 \cdot 2 + 4)$
18. $5 + 8^2 + 3^3 + 1$
19. $14 \div 7 - 3 \cdot (2 + 6)$
20. $5^2 - 2 \cdot 4 + (7 - 2)$
21. $2[12 - (2 + 2^2) \div 2]$

12.6 Problem-Solving Strategy: Working Backward

Objective: to solve word problems by working backward and other strategies

The Hawkins brothers, Sal, Al, and Cal, are taking a vacation. Sal drives one-fourth of the way. Al drives one-half of the rest of the way. They have 90 more miles to go. How far away from home are they traveling on their vacation?

You need to find how many miles away from home they are traveling. You know the end result and how much Sal and Al drove.

The end result is 90 miles. Work backward step-by-step to find the total distance from home.

They have 90 miles left to go. Since this is one-half of the remaining miles, they had 90 • 2, or 180 miles, to go before Al drove. One hundred eighty is three-fourths of the total mileage. Set up a proportion. Let m represent the number of miles they planned to go at the beginning of their trip. Solve the proportion.

$$\frac{3}{4} = \frac{180}{m}$$ left to go $(\frac{1}{2})$ Al $(\frac{1}{2})$

$3 \cdot m = 4 \cdot 180$

$3m = 720$

$m = 240$ rest of trip $(\frac{3}{4})$ Sal $(\frac{1}{4})$

They are traveling 240 miles away from home.

Start with 240 miles. Sal drove one-fourth of them, or 60 miles. There were 180 miles left to go. Al drove one-half of these, or 90 miles. They have 90 miles to go. The solution is correct.

Solve. Work backward.

1. Tina had several extra stickers that did not fit in her album. She gave $\frac{1}{4}$ to each of her three friends. She put $\frac{1}{3}$ of the remaining stickers on party invitations, and then she gave the last 8 to her sister. How many extra stickers did Tina have to start?

2. Ned has a beaker filled with water. He uses half of the water and gives half of the remaining amount to Juanita for her experiment. He has 225 mL left in the beaker. How much water was in the beaker originally?

Solve

Use any strategy to solve.

1. Diana divided her coins equally among Mary, Doug, and Nathan. Mary shared her coins equally with five other people, who each received three coins. How many coins did Diana have in the beginning?

2. Sally Wu works 3 h 15 min one day, 4 h 20 min the next day, and 3 h 30 min the third day. She earns $4.25 per hour. What does she earn for the 3 days?

3. A restaurant sold 2 million hamburgers, each 2 centimeters thick. If these hamburgers were stacked, how many kilometers high would the stack be?

4. A stack of 10 five-dollar bills is 1 millimeter high. How much money is in a stack of five-dollar bills that is 1 meter high?

5. Palindrome numbers read the same forward and backward. There was only one year in the nineteenth century that was a palindrome. What year was it?

Use the chart to solve.

6. The Rent-A-Car Company rents cars for a fixed amount each day plus a fee for each mile driven. Let x = the mileage and y = the total cost. What is the slope of a line that represents the information in the chart?

7. Sam's 5-day fee is $90.00. What is the fee per mile to the nearest cent?

8. Linda rented a car for 3 days at $15.00 per day. What is the fee per mile to the nearest cent?

	Mileage	Cost
Sam	450	$126
Linda	350	$98

Chapter 12 Review

Language and Concepts

Choose the correct term to complete each sentence.

1. The graph of a _____ function is not a straight line.
2. A _____ function is a function in which the greatest power of a variable is 2.
3. An algebraic expression with one or more monomials linked by addition or subtraction is called a _____.
4. The graph of a quadratic function is a _____.
5. A _____ is one term in a polynomial.

linear
nonlinear
quadratic
parabola
monomial
polynomial

Skills and Problem Solving

Determine whether each graph represents a *linear function* or a *nonlinear function*. Explain. Section 12.1

6.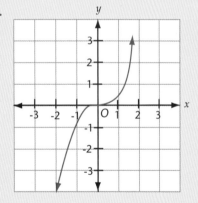

7.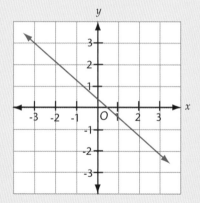

Determine whether each equation or table represents a *linear function* or a *nonlinear function*. Explain. Section 12.1

8. $y = 4x^2 - 2$

9. $y = -3x + 1$

10.
x	y
10	5
8	4
6	3
4	2

Graph each function on graph paper. Section 12.2

11. $y = x^2 - 4$
12. $y = 3x^2$
13. $y = 2x^2 - 2$

Simplify each polynomial. Section 12.3

14. $5a - 6b + 3b + 4a$
15. $8x^2 + 7x - 3x$
16. $4d + 7d^2 - 6d + 4d^2 - 9$

Add or subtract. Section 12.4

17. $(8x + 4) + (9x - 1)$
18. $(7w^2 + 2w - 5) - (4w^2 + 6w - 2)$
19. $(5r^2 - r + 1) + (r^2 + r + 1)$
20. $(6p^2 - 2p + 5) - (5p^2 + 3)$

Multiply. Section 12.5

21. $x(x + 8)$
22. $2a(3a^2 + 5a + 8)$
23. $-5w(-3w^2 - 6w - 9)$

Solve. Sections 12.2–12.6

24. Dwayne passes out pencils for a class exercise. He gives one-half of the pencils to Todd, who will help pass them out. He then gives one-third of the remainder to Sara, who will pass them down her row. Dwayne has 12 pencils left. How many pencils did he start with?

25. Write an expression in simplest form for the perimeter of the rectangle to the right. Find the perimeter of the rectangle if $x = 4$.

26. Write an expression in simplest form for the area of the rectangle to the right. Find the area of the rectangle if $x = 4$.

Chapter 12 Test

Determine whether each graph, equation, or table represents a *linear function* or a *nonlinear function*.

1. $y = -5x^2 + 3$

2.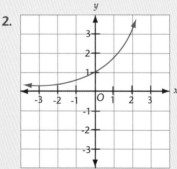

3.
x	y
2	5
4	7
6	9
8	11

Graph each function on graph paper.

4. $y = -4x^2$
5. $y = x^2 - 3$
6. $y = -3x^2 + 5$

Simplify each polynomial.

7. $-8r + 6 + 7r - 2$
8. $5x^2 - 6x + 4x - 7x^2$
9. $-3a^2 + 4a - 9a^2 + 8a$

Add or subtract.

10. $(2x^2 + 8) - (-7x^2 - 9)$
11. $(-3m^2 + 3m - 7) - (2m^2 + 5m - 4)$
12. $(-x^2 + 5x + 7) + (2x^2 - 4x - 6)$
13. $(5r^2 - 3r + 2) + (3r - 4)$

Multiply.

14. $2x(5x - 3)$
15. $4m(6m^2 + 8m - 2)$
16. $-6x(-x^2 + 3x - 1)$

Solve.

17. Susie, Maria, and Angie are playing a game of "I'm Thinking." Angie is thinking of a number between 0 and 50. The product of the number and 10, plus 7, minus 14, is 39. What is Angie's number?

Use the rectangle to the right to solve each problem.

18. Write an expression in simplest form for the perimeter of the rectangle.

19. Make a table of values, and graph the function.

20. What is the perimeter of the rectangle if $x = 5$?

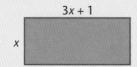

Change of Pace

Cubic Functions

A quadratic function is one example of a nonlinear function. The graph of a quadratic function is a parabola. Another example of a nonlinear function is a cubic function.

A cubic function is a function in which the greatest power of the variable is 3. You can make a table and graph a cubic function.

Graph $y = x^3$.

Find several ordered pairs that satisfy the equation.

x	$y = x^3$	y	(x, y)
-2	$y = (-2)^3$	-8	(-2, -8)
-1	$y = (-1)^3$	-1	(-1, -1)
0	$y = (0)^3$	0	(0, 0)
1	$y = (1)^3$	1	(1, 1)
2	$y = (2)^3$	8	(2, 8)

Graph the ordered pairs.

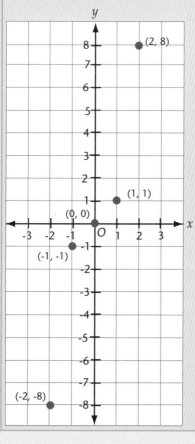

Connect the points.

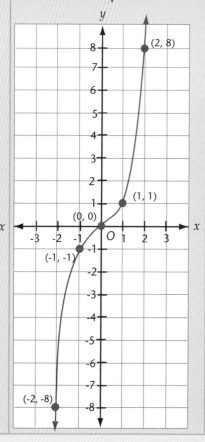

Graph each cubic function.

1. $y = 2x^3$
2. $y = x^3 + 1$
3. $y = x^3 - 2$
4. $y = 3x^3$
5. $y = -x^3$
6. $y = 2x^3 - 1$

Cumulative Test

1. $-15 \cdot -\frac{1}{3} =$ _____
   a. -45
   b. -5
   c. 5
   d. 45

2. Which number replaces n to make a true sentence?
   $100 \div n = 100$
   a. 0
   b. 0.01
   c. 1
   d. 100

3. Which point represents -1.3?

   K L   M N
   +-●●-+-●●-+
   -2 -1 0 1 2

   a. K
   b. L
   c. M
   d. N

4. $-4 + 2(-3 - 2) =$ _____
   a. -30
   b. -14
   c. -10
   d. 10

5. Evaluate $a - 2b$ if $a = 3$ and $b = \frac{3}{5}$.
   a. $1\frac{4}{5}$
   b. $2\frac{1}{5}$
   c. $2\frac{2}{5}$
   d. none of the above

6. What is the sum of the measures of the angles in a square?
   a. 90°
   b. 180°
   c. 360°
   d. none of the above

7. The chart shows how far Maura ran each day. On which day did she run the greatest distance?
   a. Thursday
   b. Saturday
   c. Monday
   d. Wednesday

Thursday	5.2 km
Saturday	4.75 km
Monday	5.6 km
Wednesday	5.18 km

8. Kenji collected some old books. Sam collected twice as many as Kenji. Which statement below contains the same information?
   a. Kenji has twice as many books as Sam.
   b. Sam has half as many books as Kenji.
   c. Kenji has half as many books as Sam.
   d. Sam has as many books as Kenji.

9. Which figures appear to be congruent?
   a. X and Y
   b. X and Z
   c. Y and Z
   d. none of the above

10. The cost of renting a car includes a charge of $18.00 per day and 20¢ per mile driven. Which of the following equations can be used to find the cost for 1 day, driving n miles?
    a. $c = 18.20n$
    b. $c = 18.20 + n$
    c. $c = 0.2 + 18n$
    d. $c = 18 + 0.2n$

Appendix

Squares and Approximate Square Roots

N	N²	√N	N	N²	√N
1	1	1.000	51	2,601	7.141
2	4	1.414	52	2,704	7.211
3	9	1.732	53	2,809	7.280
4	16	2.000	54	2,916	7.348
5	25	2.236	55	3,025	7.416
6	36	2.449	56	3,136	7.483
7	49	2.646	57	3,249	7.550
8	64	2.828	58	3,364	7.616
9	81	3.000	59	3,481	7.681
10	100	3.162	60	3,600	7.746
11	121	3.317	61	3,721	7.810
12	144	3.464	62	3,844	7.874
13	169	3.606	63	3,969	7.937
14	196	3.742	64	4,096	8.000
15	225	3.873	65	4,225	8.062
16	256	4.000	66	4,356	8.124
17	289	4.123	67	4,489	8.185
18	324	4.243	68	4,624	8.246
19	361	4.359	69	4,761	8.307
20	400	4.472	70	4,900	8.367
21	441	4.583	71	5,041	8.426
22	484	4.690	72	5,184	8.485
23	529	4.796	73	5,329	8.544
24	576	4.899	74	5,476	8.602
25	625	5.000	75	5,625	8.660
26	676	5.099	76	5,776	8.718
27	729	5.196	77	5,929	8.775
28	784	5.292	78	6,084	8.832
29	841	5.385	79	6,241	8.888
30	900	5.477	80	6,400	8.944
31	961	5.568	81	6,561	9.000
32	1,024	5.657	82	6,724	9.055
33	1,089	5.745	83	6,889	9.110
34	1,156	5.831	84	7,056	9.165
35	1,225	5.916	85	7,225	9.220
36	1,296	6.000	86	7,396	9.274
37	1,369	6.083	87	7,569	9.327
38	1,444	6.164	88	7,744	9.381
39	1,521	6.245	89	7,921	9.434
40	1,600	6.325	90	8,100	9.487
41	1,681	6.403	91	8,281	9.539
42	1,764	6.481	92	8,464	9.592
43	1,849	6.557	93	8,649	9.644
44	1,936	6.633	94	8,836	9.695
45	2,025	6.708	95	9,025	9.747
46	2,116	6.782	96	9,216	9.798
47	2,209	6.856	97	9,409	9.849
48	2,304	6.928	98	9,604	9.899
49	2,401	7.000	99	9,801	9.950
50	2,500	7.071	100	10,000	10.000

Trigonometric Ratios

Angle	sin	cos	tan	Angle	sin	cos	tan
0°	0.0000	1.0000	0.0000	45°	0.7071	0.7071	1.0000
1°	0.0175	0.9998	0.0175	46°	0.7193	0.6947	1.0355
2°	0.0349	0.9994	0.0349	47°	0.7314	0.6820	1.0724
3°	0.0523	0.9986	0.0524	48°	0.7431	0.6691	1.1106
4°	0.0698	0.9976	0.0699	49°	0.7547	0.6561	1.1504
5°	0.0872	0.9962	0.0875	50°	0.7660	0.6428	1.1918
6°	0.1045	0.9945	0.1051	51°	0.7771	0.6293	1.2349
7°	0.1219	0.9925	0.1228	52°	0.7880	0.6157	1.2799
8°	0.1392	0.9903	0.1405	53°	0.7986	0.6018	1.3270
9°	0.1564	0.9877	0.1584	54°	0.8090	0.5878	1.3764
10°	0.1736	0.9848	0.1763	55°	0.8192	0.5736	1.4281
11°	0.1908	0.9816	0.1944	56°	0.8290	0.5592	1.4826
12°	0.2079	0.9781	0.2126	57°	0.8387	0.5446	1.5399
13°	0.2250	0.9744	0.2309	58°	0.8480	0.5299	1.6003
14°	0.2419	0.9703	0.2493	59°	0.8572	0.5150	1.6643
15°	0.2588	0.9659	0.2679	60°	0.8660	0.5000	1.7321
16°	0.2756	0.9613	0.2867	61°	0.8746	0.4848	1.8040
17°	0.2924	0.9563	0.3057	62°	0.8829	0.4695	1.8807
18°	0.3090	0.9511	0.3249	63°	0.8910	0.4540	1.9626
19°	0.3256	0.9455	0.3443	64°	0.8988	0.4384	2.0503
20°	0.3420	0.9397	0.3640	65°	0.9063	0.4226	2.1445
21°	0.3584	0.9336	0.3839	66°	0.9135	0.4067	2.2460
22°	0.3746	0.9272	0.4040	67°	0.9205	0.3907	2.3559
23°	0.3907	0.9205	0.4245	68°	0.9272	0.3746	2.4751
24°	0.4067	0.9135	0.4452	69°	0.9336	0.3584	2.6051
25°	0.4226	0.9063	0.4663	70°	0.9397	0.3420	2.7475
26°	0.4384	0.8988	0.4877	71°	0.9455	0.3256	2.9042
27°	0.4540	0.8910	0.5095	72°	0.9511	0.3090	3.0777
28°	0.4695	0.8829	0.5317	73°	0.9563	0.2924	3.2709
29°	0.4848	0.8746	0.5543	74°	0.9613	0.2756	3.4874
30°	0.5000	0.8660	0.5774	75°	0.9659	0.2588	3.7321
31°	0.5150	0.8572	0.6009	76°	0.9703	0.2419	4.0108
32°	0.5299	0.8480	0.6249	77°	0.9744	0.2250	4.3315
33°	0.5446	0.8387	0.6494	78°	0.9781	0.2079	4.7046
34°	0.5592	0.8290	0.6745	79°	0.9816	0.1908	5.1446
35°	0.5736	0.8192	0.7002	80°	0.9848	0.1736	5.6713
36°	0.5878	0.8090	0.7265	81°	0.9877	0.1564	6.3138
37°	0.6018	0.7986	0.7536	82°	0.9903	0.1392	7.1154
38°	0.6157	0.7880	0.7813	83°	0.9925	0.1219	8.1443
39°	0.6293	0.7771	0.8098	84°	0.9945	0.1045	9.5144
40°	0.6428	0.7660	0.8391	85°	0.9962	0.0872	11.4301
41°	0.6561	0.7547	0.8693	86°	0.9976	0.0698	14.3007
42°	0.6691	0.7431	0.9004	87°	0.9986	0.0523	19.0811
43°	0.6820	0.7314	0.9325	88°	0.9994	0.0349	28.6363
44°	0.6947	0.7193	0.9657	89°	0.9998	0.0175	57.2900
45°	0.7071	0.7071	1.0000	90°	1.0000	0.0000	∞

Mathematical Symbols

$=$	is equal to
$\neq$	is not equal to
$>$	is greater than
$<$	is less than
$\geq$	is greater than or equal to
$\leq$	is less than or equal to
$\approx$	is approximately equal to
$\equiv$	is congruent to
$\sim$	is similar to
$\overleftrightarrow{AB}$	line AB
$\overline{AB}$	line segment AB
$\overrightarrow{AB}$	ray AB
$\triangle$	triangle
$\angle$	angle
$\perp$	is perpendicular to
$\parallel$	is parallel to
°	degrees
π	pi
$\sqrt{}$	square root
%	percent
+4	positive four
-4	negative four

Formulas

$A = lw$	area of a rectangle
$A = b \times h$	area of a parallelogram
$A = \frac{1}{2} \times b \times h$	area of a triangle
$A = \frac{1}{2} \times h \times (a + b)$	area of a trapezoid
$A = \pi r^2$	area of a circle
$C = \pi d$	circumference of a circle
$S = 4\pi r^2$	surface area of a sphere
$S = \pi r^2 + \pi rs$	surface area of a cone
$S = 2\pi r^2 + 2\pi rh$	surface area of a cylinder
$V = lwh$	volume of a rectangular prism
$V = \frac{4\pi r^3}{3}$	volume of a sphere
$V = Bh$	volume of a prism
$V = \frac{1}{3}lwh$	volume of a pyramid
$V = \pi r^2 h$	volume of a cylinder
$V = \frac{1}{3}\pi r^2 h$	volume of a cone
$I = p \times r \times t$	interest
$m = \frac{y_2 - y_1}{x_2 - x_1}$	slope of a line
$F = \frac{9}{5} \times C + 32$	changing Celsius to Fahrenheit
$C = \frac{5}{9} \times (F - 32)$	changing Fahrenheit to Celsius

Metric System of Measurement

Prefixes
- kilo (k) = thousand
- hecto (h) = hundred
- deka (da) = ten
- deci (d) = tenth
- centi (c) = hundredth
- milli (m) = thousandth

Length
- 1 centimeter (cm) = 10 millimeters (mm)
- 1 meter (m) = 100 centimeters or 1,000 millimeters
- 1 kilometer (km) = 1,000 meters

Mass
- 1 gram (g) = 1,000 milligrams (mg)
- 1 kilogram (kg) = 1,000 grams
- 1 metric ton (T) = 1,000 kilograms

Capacity
- 1 liter (L) = 1,000 milliliters (mL)
- 1 kiloliter (kL) = 1,000 liters

Customary System of Measurement

Length
- 1 foot (ft) = 12 inches (in.)
- 1 yard (yd) = 3 feet or 36 inches
- 1 mile (mi) = 1,760 yards or 5,280 feet

Weight
- 1 pound (lb) = 16 ounces (oz)
- 1 ton = 2,000 pounds

Capacity
- 1 cup = 8 fluid ounces (fl oz)
- 1 pint (pt) = 2 cups
- 1 quart (qt) = 2 pints
- 1 gallon (gal) = 4 quarts

Glossary

A

absolute value 1.5 The number of units a number is from 0 on the number line. The symbol for absolute value is two vertical lines.

$$|-4| = 4$$

acute angle 6.7 An angle that has a degree measure between 0° and 90°.

acute triangle 6.7 A triangle with three acute angles.

addition property of equality 1.9 Adding the same number to each side of an equation results in an equivalent equation.

$$m - 3 = 8$$
$$m - 3 + 3 = 8 + 3$$
$$m = 11$$

adjacent angles 6.1 Angles that have a common side, the same vertex, and do not overlap. ∠CDW and ∠WDB are adjacent.

algebraic expression 1.3 A phrase that contains numbers, variables, and operation symbols.

$$a - 7$$

alternate exterior angles 6.2 When a transversal intersects two lines, the angles outside the two lines and on opposite sides of the transversal are alternate exterior angles. The alternate exterior angles below are ∠1 and ∠7, and ∠2 and ∠8.

alternate interior angles 6.2 When a transversal intersects two lines, the angles between the two lines and on opposite sides of the transversal are alternate interior angles. The alternate interior angles below are ∠3 and ∠5, and ∠4 and ∠6.

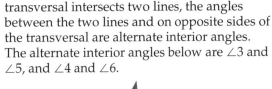

angle 6.1 Two rays with a common endpoint. Angle DEF is symbolized as ∠DEF.

area 7.1 The number of square units that cover a surface.

arithmetic sequence 11.1 A list of numbers where consecutive numbers have the same difference.

$$20, 35, 50, 65$$
$$+15 +15 +15$$

associative property 1.4 The way in which addends or factors are grouped does *not* change the answer.

$$(20 + 30) + 40 = 20 + (30 + 40)$$
$$(6 \cdot 7) \cdot 3 = 6 \cdot (7 \cdot 3)$$

B

bar graph 8.3 A graph using bars to show comparisons between different values.

base (of an exponent) 2.7 The number that is used as a factor in a power. In 4^3, the base is 4.

bases (of a prism) 7.3 Two parallel congruent faces.

biased sample 8.1 A sample in which certain parts of the population are favored.

bisect 6.4 To separate a figure into two congruent parts.

box-and-whisker plot 8.6 A way to organize numerical data to show the median, quartiles, and interquartile range.

C

center 7.2 The middle point of a circle.

chord 7.2 A line segment with both endpoints on a circle.

circle 7.2 A plane figure formed by all points that are the same distance from a given point in the plane called the *center*.

circle graph 8.2 Compares parts of a whole by showing the percent each part is of the whole.

circumference (C) 7.2 The distance around the circle.

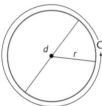

combination 9.4 A selection of items in which order is not important.

common difference 11.1 The difference between two consecutive terms in an arithmetic sequence.

common ratio 11.1 The quotient between any two consecutive numbers is the same in a geometric sequence.

commutative property 1.4 The order in which numbers are added or multiplied does not change the answer.

$$8 + 5 = 5 + 8$$
$$2 \cdot 30 = 30 \cdot 2$$

complementary angles 6.1 Two angles are complementary if the sum of their degree measures is 90°. ∠ADB and ∠BDC are complementary.

compound event 9.7 Two or more simple events.

cone 7.3 A solid with a circular base and one vertex.

congruent figures 6.9 Figures that have the same size and shape.

constructions 6.3 Drawings made with a straightedge and a compass.

convenience sample 8.1 A type of biased sample that includes members of the population that are easily accessible.

conversion factors 4.4 Rates equal to 1.

correlation 8.9 A relationship between two variables.

corresponding angles 6.2 When a transversal intersects two lines, the angles in similar positions are called corresponding angles. The corresponding angles below are ∠1 and ∠5, ∠2 and ∠6, ∠3 and ∠7, and ∠4 and ∠8.

cosine (cos) 4.11 The ratio of the measure of the side adjacent to an acute angle in a right triangle to the measure of the hypotenuse.

cross products 4.5 In the proportion $\frac{3}{6} = \frac{4}{8}$, the cross products are 3×8 and 6×4.

cylinder 7.3
A solid that has two parallel and congruent circular bases.

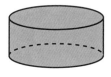

D

data 8.2 Numerical information.

dependent events 9.7 Two events in which the outcome of the second event is affected by the outcome of the first event.

diameter 7.2 A chord of a circle that contains the center of the circle.

dilation 4.8 An image that has been altered in size but not shape.

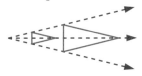

dimensional analysis 4.4 A way to convert from one unit to another unit using a conversion factor.

discount 5.6 The amount of reduction from the regular price.

distributive property 1.8 The product of a number and a sum is equal to the sum of the products.

$$2 \cdot (4 + 3) = (2 \cdot 4) + (2 \cdot 3)$$

division property of equality 1.9 Dividing each side of an equation by the same nonzero number results in an equivalent equation.

$$6n = 78$$
$$\frac{6n}{6} = \frac{78}{6}$$
$$n = 13$$

domain (of a relation) 11.2 The set of all first components in a set of ordered pairs.

E

equation 1.3 A mathematical sentence with an equals sign.

$$25 - 17 = 8 \qquad c + 3 = 9$$

equilateral triangle 6.7 A triangle that has three congruent sides.

equivalent equation 1.9 An equation that has the same solution.

event 9.1 A specific outcome.

experimental probability 9.6 The probability based on the results of an experiment.

exponent 2.7 A number used to tell how many times the base is used as a factor. In 10^2, the exponent is 2.

F

factorial 9.3 The product of all the whole numbers, except 0, that are less than or equal to that number.

$$3! = 3 \cdot 2 \cdot 1$$

formula 1.10 A mathematical statement that shows the relationship between certain quantities.

function 11.2 A relationship where the output depends on the input, and there is exactly one output value for each input value.

fundamental counting principle 9.2 All possible outcomes in a sample space can be found by multiplying the number of ways each event can occur.

G

geometric sequence 11.1 A list of numbers where each term is formed by multiplying the previous term by the same number.

$$2, 4, 8, 16, 32, \ldots$$

golden rectangle 6.6 The length of a golden rectangle divided by its width is about 1.6.

H

histogram 8.7 A vertical bar graph with bars next to each other. A histogram is used to show the frequency of numbers or items in a set of data.

hypotenuse 3.3 The side opposite the right angle in a right triangle.

I

identity property of addition 1.4 When 0 is added to a number, the sum is that number.

$$213 + 0 = 213$$

identity property of multiplication 1.4 When a number is multiplied by 1, the product is that number.

$$65 \cdot 1 = 65$$

independent events 9.7 Two events in which the outcome of the second event is not affected by the outcome of the first event.

indirect measurement 4.10 A method of measuring distances by solving proportions.

inequality 10.5 A mathematical sentence that contains a symbol, such as $<, >, \leq, \geq,$ or $\neq$.

$$7 < 8 \quad 14 \geq x - 1.5 \quad 0.5 \neq \frac{1}{3}$$

integers 1.5, 3.2 The whole numbers and their opposites.

$$\ldots, -3, -2, -1, 0, 1, 2, 3, \ldots$$

interest 5.7 The amount of money you pay on a loan or earn on an investment.

interquartile 8.5 The middle half of the data.

interquartile range 8.5 The difference between the lower and upper quartiles of the data.

inverse property of addition 1.8 When a number is added to its opposite, the sum is 0.

inverse property of multiplication 1.8 When a number is multiplied by its reciprocal, the product is 1.

irrational number 3.2 A number that can be named by a nonterminating, nonrepeating decimal.

isosceles triangle 6.7 A triangle that has at least two congruent sides.

L

like terms 10.2 Terms that have the same variables.

line 6.2 All the points in a never-ending straight path. Line ST is symbolized as $\overleftrightarrow{ST}$.

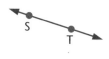

line graph 8.7 A graph in which line segments are used to compare changes in data over time.

line plot 8.7 A picture of information on a number line. An **x** is often used to indicate the occurrence of a number.

line of symmetry 6.11 The line drawn through a figure so that the figure on one side is a mirror image of the figure on the other side.

line segment 6.3 Two endpoints and the straight path between them. Line segment PQ is symbolized as $\overline{PQ}$.

linear function 11.3 A function in which the graph of the solutions forms a line.

M

markup 5.6 The percent of increase in price to cover expenses and make a profit.

mean 8.3 The sum of the numbers divided by the number of addends of a data set.

measure of central tendency 8.3 A number that represents the center of a set of data as in the *mean*, *median*, and *mode*.

measure of variation 8.5 The spread of values in a set of data.

median 8.3 The middle number in an ordered set of data. When there are two middle numbers, the median is the mean of those two numbers.

mode 8.3 The number that appears most often in a set of data.

monomials 12.3 The terms in an expression.

multiplication property of equality 1.9 Multiplying each side of an equation by the same number results in an equivalent equation.

$$\frac{x}{7} = 9$$
$$\frac{x}{7} \cdot 7 = 9 \cdot 7$$
$$x = 63$$

multiplicative inverses 2.4 Two numbers whose product is 1. $\frac{3}{4}$ and $\frac{4}{3}$ are multiplicative inverses of each other.

mutually exclusive events 9.1 Two events that cannot happen at the same time.

N

negative correlation 8.9 The relationship in which one variable tends to increase as the other tends to decrease.

negative integer 1.5 Any number to the left of 0 on a number line.

-100   -8   -5

nonlinear function 12.1 A function that is not a straight line when graphed because it does not have a constant rate of change.

O

obtuse angle 6.7 An angle that has a degree measure between 90° and 180°.

obtuse triangle 6.7 A triangle with one obtuse angle.

odds against 9.8 The ratio of the number of ways an event cannot occur to the number of ways the event can occur.

odds in favor 9.8 The ratio of the number of ways an event can occur to the number of ways the event cannot occur.

opposites 1.5 Two numbers that are the same distance from 0 on the number line. The opposite of 7 is -7 and vice versa.

outlier 8.5 A value that is 1.5 times the value of the interquartile range.

P

parabola 12.2 The curve of the graph of a quadratic function.

parallel lines 6.2 Lines in the same plane that do not intersect.

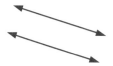

parallelogram 6.7 A quadrilateral with two pairs of parallel sides.

percent 5.1 An expression of hundredths using the percent symbol (%).

$$23\% = \frac{23}{100} = 0.23$$

percent of change 5.5 The percent that a quantity increases or decreases from its original amount.

perfect square 3.1 The square of an integer. The number 64 is a perfect square, since $64 = 8 \times 8$.

permutation 9.3 An arrangement in which order is important.

perpendicular lines 6.5
Two lines that intersect to form right angles.

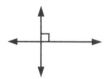

pi (π) 7.2 The ratio of the circumference of a circle to its diameter. An approximation for π is 3.14 or $\frac{22}{7}$.

pictograph 8.7 A graph that uses pictures to display data.

polygon 6.8 A closed plane figure formed by line segments.

polyhedron 7.3 A solid with flat surfaces called *faces*.

polynomial 12.3 An algebraic expression with one or more monomials linked by addition or subtraction.

positive correlation 8.9 A relationship in which the values of two variables increase or decrease together.

positive integer 1.5 Any number to the right of 0 on a number line.

5   6   20

power 2.7 A number that is expressed using an exponent.

principal 5.7 The amount you borrow or invest.

prism 7.3 A polyhedron with two parallel congruent bases that are shaped like polygons. The other faces of the prism are shaped like parallelograms.

probability 9.1 The ratio of the number of ways an event can occur to the total number of possible outcomes.

proportion 4.5 An equation that states two ratios are equivalent.

$$\frac{5}{6} = \frac{10}{12}$$

pyramid 7.3 A polyhedron with one base that is shaped like a polygon. The other faces of the pyramid are triangular.

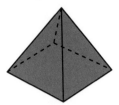

Pythagorean theorem 3.3 In a right triangle, the square of the length of the hypotenuse is equal to the sum of the squares of the lengths of the other two sides.

Q

quadratic function 12.2 A function in which the greatest power of the variable is 2.

quadrilateral 6.7 A polygon with four sides.

quartiles 8.5 Four equal parts into which an ordered set of data can be divided.

R

radius 7.2 A line segment whose endpoints are the center of a circle and a point on the circle.

range 8.5 The difference between the greatest number and the least number in a set of data.

range (of a relation) 11.2 The set of all second components in a set of ordered pairs.

rate 4.1 A ratio of two measurements having different units.

$$\frac{5 \text{ km}}{2 \text{ min}} \quad 8 \text{ mph}$$

rate (of interest) 5.7 A percent that must be converted to a decimal.

rate of change 4.2 Describes how one quantity changes in relation to another.

ratio 4.1 A comparison of two numbers. The ratio of 2 to 3 can be written as 2 out 3, 2 to 3, 2:3, or $\frac{2}{3}$.

rational number 2.1, 3.2 Any number that can be expressed as a quotient of two integers when the divisor is not 0.

real number 3.2 Either a rational number or an irrational number.

reciprocals 1.8, 2.4 Two numbers whose product is 1. Since $\frac{3}{4} \times \frac{4}{3} = 1$, $\frac{3}{4}$ and $\frac{4}{3}$ are reciprocals of each other. Reciprocals are also called *multiplicative inverses*.

rectangle 6.7 A parallelogram with four congruent angles.

rectangular prism 7.4 A polyhedron that has six rectangular sides.

reflection 6.11 A figure that is the mirror image of another figure caused by flipping it over a line.

regular polygon 6.8 A polygon in which all sides are congruent and all angles are congruent.

repeating decimal 2.1 A decimal whose digits repeat in groups of one or more. The decimals equivalent to $\frac{2}{3}$ and $\frac{5}{11}$ are repeating decimals.

$$\frac{2}{3} = 0.6666\ldots \text{ or } 0.\overline{6}$$

$$\frac{5}{11} = 0.4545\ldots \text{ or } 0.\overline{45}$$

rhombus 6.7 A parallelogram with four congruent sides.

right angle 6.1 An angle that has a degree measure of 90°.

right triangle 6.7 A triangle that has a right angle.

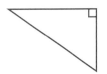

rise 4.3, 11.5 The change in y found by subtracting the y-coordinates.

rotation 6.10 A transformation that turns each point the same number of degrees around a common point.

run 4.3, 11.5 The change in x found by subtracting the x-coordinates.

S

sale price 5.6 The regular price minus the discount.

sample 8.1 A small group chosen at random from a larger group. Predictions about the larger group are made by studying the sample.

sample space 9.1 A list of all possible outcomes.

scale drawing 4.9 A proportional representation of something that is too large or small to be conveniently drawn in actual size.

scale factor 4.8 The ratio found by comparing distances from the center of a dilation to the original image's center.

scalene triangle 6.7 A triangle that has no congruent sides.

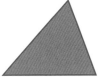

scatter plot 8.9 A graph of two variables.

scientific notation 2.10 A way of expressing numbers as the product of a number between 1 and 10 and a power of 10.

$$1{,}017{,}000 = 1.017 \cdot 10^6$$

sequence 11.1 A list of numbers in a certain order.

arithmetic sequence: 0, 2, 4, 6, 8, . . .
geometric sequence: 1, 2, 4, 8, 16, . . .

similar figures 4.7 Figures that have the same shape and proportional sides.

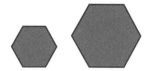

simple random sample 8.1 A sample where each person or object has an equally likely chance of being chosen.

simplest form 2.3 A fraction is in simplest form when the greatest common factor of the numerator and denominator is 1.

sine (sin) 4.11 The ratio of the measure of the side opposite an acute angle in a right triangle to the measure of the hypotenuse.

slope 4.3, 11.5 The slope of a line is the vertical change (change in y) divided by the horizontal change (change in x).

slope-intercept form 11.6 An equation written in the form of $y = mx + b$.

solution (of an equation) 1.9 A replacement for the variable that makes an equation true. The solution of the equation $y + 2 = 7$ is 5.

solution set 10.5 The set of all replacements for the variable that makes the inequality true.

solve 1.9 To find the number(s) that make an equation or inequality true.

square 6.7 A parallelogram with four congruent sides and four congruent angles.

square root 3.1 One of the two equal factors of a number. 7 is the square root of 49 since $49 = 7 \times 7$. -7 is another square root of 49.

$$\sqrt{49} = 7$$

statistical data 8.1 Pieces of information that are collected, analyzed, and presented.

stem-and-leaf plot 8.4 A way to organize numerical data. The greatest place value of the data is used for the stem. The next greatest place value is used for the leaves.

stratified random sample 8.1 A simple random sample where the population is divided into smaller groups with certain characteristics.

subtraction property of equality 1.9 Subtracting the same number from each side of an equation results in an equivalent equation.

$$\begin{aligned} x + 9 &= 13 \\ x + 9 - 9 &= 13 - 9 \\ x &= 4 \end{aligned}$$

supplementary angles 6.1 Two angles are supplementary if the sum of their degree measures is 180°. $\angle ADB$ and $\angle BDC$ are supplementary.

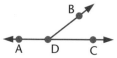

surface area 7.6 The sum of the areas of the surfaces of a solid.

systematic random sample 8.1 A random sample where people or objects are selected according to time intervals.

T

table 8.7 A way to list data individually or by groups.

tangent (tan) 4.11 The ratio of the measure of the side opposite an acute angle in a right triangle to the measure of the side adjacent.

term 10.2 A number, a variable, or the product of a number and a variable.

terminating decimal 2.1 A decimal whose digits end. The decimals equivalent to $\frac{1}{4}$ and $\frac{3}{8}$ are terminating decimals.

$$\frac{1}{4} = 0.25 \qquad \frac{3}{8} = 0.375$$

theoretical probability 9.6 Probability based on characteristics or known facts.

three-dimensional figure 7.3 An object that has length, width, and height. Also called a *solid*.

translation 6.10 A transformation that moves each point on a figure the same distance and in the same direction.

transversal 6.2 A line that intersects two or more lines. $\overleftrightarrow{JK}$ is a transversal.

trapezoid 6.7 A quadrilateral with exactly one pair of parallel sides.

tree diagram 9.2 A diagram that shows the possible outcomes of an event.

triangle 6.7 A polygon with three sides.

triangular prism 7.4 A prism with triangles for bases.

U

unbiased sample 8.1 A random sample in which each person or object has an equally likely chance of being selected.

unit rate 4.1 A rate with a denominator of 1.

V

variable 1.2 A symbol, usually a letter, that is used to stand for some number. In the expression $d + 3$, the variable is d.

vertex 6.1 The common endpoint of the rays forming an angle. The point where line segments meet in a polygon.

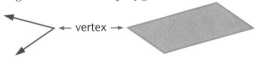

vertical angles 6.1 Angles formed by two intersecting lines. $\angle 1$ and $\angle 3$ are vertical angles. $\angle 2$ and $\angle 4$ are vertical angles.

volume 7.4 The amount of space that a solid contains.

voluntary response sample 8.1 A sample that includes only those who want to participate.

X

x-axis 11.3 The horizontal number line in a coordinate system.

x-intercept 11.3 The value of x where the graph crosses the x-axis.

Y

y-axis 11.3 The vertical number line in a coordinate system.

y-intercept 11.3 The value of y where the graph crosses the y-axis.

Index

A

Absolute value, 12–13
Absolute value bars (||), 12–13
Acting it out, 250–251
Acute angle, 158–159
Acute triangle, 158–159
Addition
   of algebraic fractions, 40
   associative property of, 8–9
   commutative property of, 8–9
   of fractions, 38–40
   identity property of, 8–9
   of inequalities, 268–269
   of integers, 14–15
   inverse property of, 20–21
   of mixed numbers, 38–40
   order of operations and, 2–3
   of polynomials, 314–315
   of probabilities, 232–233
Addition property of equality, 22–23
Adjacent angles, 144–145
Algebraic expressions, 6–7
Algebraic fractions, 40, 44
Algorithm, Euclidean, 61
Alternate exterior angles, 146–148
Alternate interior angles, 146–148
Angles, 144–145, 148
   acute, 158–159
   adjacent, 144–145
   alternate exterior, 146–148
   alternate interior, 146–148
   bisecting, 152–153
   complementary, 144–145
   congruent, 150–151, 162–163
   corresponding, 146–148
   cosine of, 110–111, 112–113
   of elevation, 117
   obtuse, 158–159
   of polygons, 160–161
   right, 144–145
   sine of, 110–111, 112–113
   sum of, in triangle, 144–145
   supplementary, 144–145
   tangent of, 110–111, 112–113
   vertical, 144–145
Applications. *See* Problem solving.
Approximately equal to symbol ($\approx$), 110–111
Area
   of circle, 182–184
   of parallelogram, 178–180
   surface
      of cone, 194–195
      of cylinder, 194–195
      of prism, 192–193
      of pyramid, 192–193
      of sphere, 201
   of trapezoid, 178–180
   of triangle, 178–180
Arithmetic, modular, 31
Arithmetic sequence, 282–283
Associative property
   of addition, 8–9
   of multiplication, 8–9
Average, 208–209
Axes, xxiii
   x-axis, xxiii, 286–287
   y-axis, xxiii, 286–287

B

Bar graphs, 208–209
Base
   of an exponent, 50–51
   of a prism, 186–187
Base-two numeration system, 279
Biased sample, 204–205
Binary numeration system, 279
Bisectors
   of angles, 152–153
   of line segments, 152–153
Box-and-whisker plots, 216–217
   extremes on, 216–217
   interquartile range of, 216–217
   median on, 216–217
   quartiles on, 216–217

Brackets
  order of operations and, 2–3

C

Center, 182–184
Central tendency, 208–209
Chord, 182–184
Circle
  area of, 182–184
  center of, 182–184
  chord, 182–184
  circumference of, 182–184
  diameter of, 182–184
  radius of, 182–184
Circle graphs, 206–207
Circumference, 182–184
Combinations, 238–239
Common difference, 282–283
Common factor, xxi
  greatest, xxi
Common multiples, xxii
  least, xxii
Common ratio, 282–283
Commutative property
  of addition, 8–9
  of multiplication, 8–9
Compass, 150–151, 152–153, 154–155, 156–157
Complementary angles, 144–145
Composite number, xx
Compound event, 246–247
Cones, 186–187
  surface area of, 194–195
  volume of, 190–191
Congruent angles, 150–151, 162–163
Congruent figures, 162–163
Congruent figures symbol (≅), 162–163
Congruent line segments, 150–151
Congruent triangles, 162–163
Constructions, 150–157
  bisecting angles, 152–153
  bisecting line segments, 152–153
  congruent angles, 150–151
  congruent line segments, 150–151
  golden rectangle, 156–157
  parallel lines, 154–155
  perpendicular lines, 154–155
Convenience sample, 204–205
Coordinate plane, xxiii

Correlation, 222–223
  negative, 222–223
  positive, 222–223
Corresponding angles, 146–147
Cosine ratio, 110–111, 112–113
Cross products, 96–97, 124–125
Cubic centimeter (cm^3), 188–189
Cubic functions, 323
  graphing, 323
Cumulative frequency histograms, 229
Cumulative relative frequency, 229
Cylinders, 186–187
  surface area of, 194–195
  volume of, 188–189

D

Data, 206–207
Decagon, 160–161
Decimal point (.), xvi, 120–121
Decimals
  comparing, xvi, 36–37
  converting
    to fractions, 34–35
    fractions to, 34–35
    to mixed numbers, 34–35
    mixed numbers to, 34–35
    to percents, 120–121
    percents to, 120–121
  in expanded form, 57
  ordering, xvi, 36–37
  repeating, 34–35
  terminating, 34–35
Dependent events, 246–247
Diagrams, 196–197
Diameter, 182–184
Dilation image, 104–105
Dilations, 104–105
Discounts, 132–133
Distributive property, 20–21
Divisibility of numbers, xix
  by five (5), xix
  by four (4), xix
  by nine (9), xix
  by six (6), xix
  by ten (10), xix
  by three (3), xix
  by two (2), xix
Division
  of algebraic fractions, 44
  of fractions, 42–44

of integers, 16–18
of mixed numbers, 42–44
order of operations and, 2–3
Division property of equality, 22–23
Dodecagon, 160–161
Domain, 284–285
Drawing solid figures, 186–187

E

Egyptian mathematics, 141
Equality
    addition property of, 22–23
    division property of, 22–23
    multiplication property of, 22–23
    subtraction property of, 22–23
Equations, 6–7
    graphing, 288–289
    involving fractions and mixed numbers, 46–47
    multi-step, 262–263
    one-step, 22–23
    second-degree, 297
    translating sentences into, 6–7, 26–27
    two-step, 24–25, 258–259
    variables on both sides of, 264–265
Equilateral triangle, 158–159
Euclid, 61
Euclidean algorithm, 61
Events, 232–233
    compound, 246–247
    dependent, 246–247
    independent, 246–247
    mutually exclusive, 232–233
Expanded form, 57
    decimals in, 57
Experimental probability, 242–243
Exponential form, 50–51
Exponents, 50–51
    integers as, 54–55
    one as (1), 54–55
    negative, 54–55
    zero as (0), 54–55
Expressions
    algebraic, 6–7
    translating phrases into, 6–7
Extremes, 216–217
    on box-and-whisker plots, 216–217

F

Factored form, 50–51
Factorials, 236–237
Factors, xix
    common, xxi
    greatest common, xxi
    prime, xx
Fair game, 255
Figures
    congruent, 162–163
    similar, 100–102
    solid, drawing, 186–187
Formulas
    area of circle, 182–184
    area of parallelogram, 178–180
    area of trapezoid, 178–180
    area of triangle, 178–180
    changing Celsius to Fahrenheit, 24–25
    changing Fahrenheit to Celsius, 24–25
    circumference, 182–184
    cumulative relative frequency, 229
    distance to a horizon line, 65
    interest, 134–135
    measure of one angle of a polygon, 160–161
    percent, 124–125, 126–127
    percent of change, 130–131
    percent of markup, 132–133
    sum of angle measures of a polygon, 160–161
    surface area of cone, 194–195
    surface area of cylinder, 194–195
    surface area of prism, 192–193
    surface area of pyramid, 192–193
    surface area of sphere, 201
    volume of cone, 190–191
    volume of cylinder, 188–189
    volume of prism, 188–189
    volume of pyramid, 190–191
    volume of sphere, 201
Four-step plan, 26–27
Fractions. *See also* Ratios.
    addition of, 38–40
    algebraic, 40, 44
    comparing, 36–37
    converting
        to decimals, 34–35
        decimals to, 34–35
        to percents, 120–121
        percents to, 120–121
    division of, 42–44

equations involving, 46–47
multiplication of, 42–44
ordering, 36–37
subtraction of, 38–40
Frequency, cumulative relative, 229
Functions, 284–285
    cubic, 323
    linear, 286–287
    quadratic, 308–309
Fundamental counting principle, 234–235

G

Games
    fair, 255
    unfair, 255
Geometric sequence, 282–283
Golden rectangle, 156–157
Graphing
    applications involving, 298–299
    cubic functions, 323
    inequalities, 296–297
        on number lines, 268–269
    linear equations, 286–287, 288–289
    quadratic functions, 308–309
Graphs
    bar, 208–209
    circle, 206–207
    misleading, 220–221
Greater than or equal to symbol ($\geq$), 268–269
Greater than symbol ($>$), 268–269
Greatest common factor (GCF), xxi
Guess and check, 136–137

H

Height, 100–102, 108–109, 117
Heptagon, 160–161
Hexagon, 160–161
Hieroglyphics, 141
Histograms
    cumulative frequency, 229
Hypotenuse, 68–69
Hypsometer, 117

I

Identity property
    of addition, 8–9
    of multiplication, 8–9

Image, dilation, 104–105
Independent events, 246–247
Indirect measurement, 108–109
Inequalities
    addition of, 268–269
    applications involving, 274–275
    division of, 270–271
    graphing, 268–269, 296–297
    multiplication of, 270–271
    solving, 268–269, 270–271
    solving two-step, 272–273
    subtraction of, 268–269
Integers, 12–13, 66–67
    adding, 14–15
    as exponents, 54–55
    dividing, 16–18
    multiplying, 16–18
    negative, 12–13
    positive, 12–13
    subtracting, 14–15
Interest, 134–135
Interquartile, 212–213
Interquartile range, 212–213
    on box-and-whisker plots, 216–217
Inverse, multiplicative, 42–44
Inverse property
    of addition, 20–21
    of multiplication, 20–21
Irrational numbers, 66–67
Isosceles triangle, 158–159

L

Least common multiple (LCM), xxii
Less than or equal to symbol ($\leq$), 268–269
Less than symbol ($<$), 268–269
Like terms, combining, 260–261
Linear equations, graphing, 286–287, 288–289
Linear functions, 286–287
Line of symmetry, 168–169
Lines, 146–148
    parallel, 146–148
        constructing, 154–155
    perpendicular, 154–155
        constructing, 154–155
Line segments, 150–151
    bisecting, 152–153
    congruent, 150–151
Logic, 83
Logical thinking, 166, 193, 251

M

Magic squares, 315
Markup, 132–133
Mean, 208–209
Mean variation, 213
Measurement
 indirect, 108–109
Measures of central tendency, 208–209
Measures of variation, 212–213
Median, 208–209
 on box-and-whisker plots, 216–217
Mental math, 125
Misleading graphs, 220–221
Mixed numbers
 addition of, 38–40
 converting
  to decimals, 34–35
  decimals to, 34–35
 division of, 42–44
 equations involving, 46–47
 multiplication of, 42–44
 subtraction of, 38–40
Mode, 208–209
Modular arithmetic, 31
Monomial, 310–311
Multiples
 common, xxii
 least common, xxii
Multiplication
 associative property of, 8–9
 commutative property of, 8–9
 of fractions, 42–44
 identity property of, 8–9
 of integers, 16–18
 inverse property of, 20–21
 of mixed numbers, 42–44
 order of operations and, 2–3
 of polynomials, 316–317
 of probabilities, 234–235, 236–237, 238–239
Multiplication property of equality, 22–23
Multiplicative inverse, 42–44
Mutually exclusive events, 232–233

N

Negative correlation, 222–223
Negative integer, 12–13
Nonagon, 160–161
Not equal to symbol ($\neq$), 268–269

Number lines
 addition of integers on, 14–15
 integers on, 12–13
 multiplication of integers on, 16–18
 solutions to inequalities on, 268–269
 subtraction of integers on, 14–15
Numbers
 composite, xx
 as exponents, 50–51
 integers, 12–13
 irrational, 66–67
 mixed. *See* Mixed numbers.
 rational. *See* Rational numbers.
 real, 66–67
 square, 3, 64–65
 triangular, 3
Numeration system
 base-two, 279
 binary, 279

O

Obtuse angle, 158–159
Obtuse triangle, 158–159
Octagon, 160–161
Odds, finding, 248–249
One (1)
 as exponent, 54–55
 multiplication by, 8–9
One-step equations, 22–23
Opposites, 12–13
Order, importance of, 236–237, 238–239
Ordered pair, xxiii
Order of operations, 2–3
Origin, xxiii
Outlier, 212–213

P

Pair, ordered, xxiii
Parabola, 308–309
Parallel lines, 146–148
 constructing, 154–155
Parallelogram, 158–159
 area of, 178–180
Parentheses
 order of operations and, 2–3
Pascal, Blaise, 240–241
Pascal's triangle, 240–241
Patterns, 3, 224–225

Pentagon, 160–161
Pentominoes, 214
Percentile, 223
Percents
   applications involving, 136–137, 206–207
   and business, 132–133
   of change, 130–131
   computing mentally, 122–123
   converting
      to decimals, 120–121
      decimals to, 120–121
      to fractions, 120–121
      fractions to, 120–121
   of decrease, 130–131
   and discounts, 132–133
   identifying rate, 134–135
   of increase, 130–131, 132–133
   and interest, 134–135
   proportions and, 124–125
Permutations, 236–237
Perpendicular lines, 154–155
   constructing, 154–155
Pi (π), 182–184
Plots
   box-and-whisker, 216–217
   scatter, 222–223
   stem-and-leaf, 210–211
Polygons, 160–161
   regular, 160–161
Polyhedrons, 186–187
   surface area of, 192–193, 194–195
   volume of, 188–189, 190–191
Polynomials, 310–311
   addition of, 314–315
   multiplication of, 316–317
   subtraction of, 314–315
Positive correlation, 222–223
Positive integer, 12–13
Power, 50–51
Price, 132–133
Prime factorization, xx, xxi
Prime number, xx
Principal, 134–135
Prisms, 186–187
   rectangular, 188–189
   surface area of, 192–193
   triangular, 188–189
   volume of, 188–189
Probabilities, 232–233
   addition of, 232–233
   experimental, 242–243
   multiplication of, 234–235, 236–237, 238–239
   theoretical, 242–243
Problem solving
   acting it out, 250–251
   constructing golden rectangles, 156–157
   cosine ratio, 112–113
   diagrams, 196–197
   four-step plan, 26–27
   graphs, 298–299
   guess and check, 136–137
   identifying necessary facts, 170–171
   indirect measurement, 108–109
   inequalities, 274–275
   logical reasoning, 78–79
   patterns, 224–225
   pentominoes, 214
   proportions, 98–99
   simpler problem, 48–49
   sine ratio, 112–113
   tangent ratio, 112–113
   working backward, 318–319
Products, cross, 96–97, 124–125
Proportions, 96–97
   applications involving, 98–99
   percents and, 124–125
Protractor, 117
Pyramids, 186–187
   surface area of, 192–193
   volume of, 190–191
Pythagorean theorem, 68–69
Pythagorean triples, 77

Q

Quadratic function, 308–309
Quadrilaterals, 158–159, 160–161
Quartiles, 212–213
   on box-and-whisker plots, 216–217

R

Radius, 182–184
Range, 212–213
   on box-and-whisker plots, 216–217
Range of relations, 284–285
Rates, 86–87
   of change, 88–89
   as percents, 134–135
   unit, 86–87

Rational numbers, 34–35, 66–67
    comparing, 36–37
    ordering, 36–37
Ratios, 86–87. *See also* Percents.
    cosine, 110–111, 112–113
    sine, 110–111, 112–113
    tangent, 110–111, 112–113
Real numbers, 66–67
Reciprocals, 20–21, 42–43
Rectangle, 158–159
    golden, 156–157
Rectangular prism, 188–189
Reflections, 168–169
Regular polygons, 160–161
Repeating decimals, 34–35
Rhombus, 158–159
Right angle, 144–145
Right angle symbol (⌐), 144–145
Right triangles, 158–159. *See also* Cosine ratio; Pythagorean theorem; Sine ratio; Tangent ratio.
    30°–60°, 74–75
    45°–45°, 76–77
Rise, 90–92, 292–293
Roots, 64–65
Rotations, 164–166
Rule of 78, 303
Run, 90–92, 292–293

S

Sale price, 132–133
Sales tax, 132–133
Samples, 204–205
    biased, 204–205
    convenience, 204–205
    simple random, 204–205
    stratified random, 204–205
    systematic random, 204–205
    unbiased, 204–205
    voluntary response, 204–205
Scale drawings, 106–107
Scale factor, 104–105
Scalene triangle, 158–159
Scatter plots, 222–223
Scientific notation, 56–57
Second-degree equations, 297
Sequences, 282–283
    arithmetic, 282–283
    geometric, 282–283

Similar figures, 100–102
Similar figures symbol (~), 100–102
Simple random sample, 204–205
Sine ratio, 110–111, 112–113
Slope, 90–92, 292–293
Slope-intercept form, 294–295
Solid figures, drawing, 186–187
Solution set, 268–269
Spheres, 201
    surface area of, 201
    volume of, 201
Square, 158–159
    magic, 315
Square numbers, 3, 64–65
Square root, 64–65
Stadia, 102
Statements, 83
    conditional, 83
    truth value of, 83
Statistics
    box-and-whisker plots, 216–217
    central tendency, 208–209
    measures of variation, 212–213
    misleading graphs, 220–221
    scatter plots, 222–223
    stem-and-leaf plots, 210–211
Stem-and-leaf plots, 210–211
Stratified random sample, 204–205
Subtraction
    of fractions, 38–40
    of inequalities, 268–269
    of integers, 14–15
    of mixed numbers, 38–40
    order of operations and, 2–3
    of polynomials, 314–315
Subtraction property of equality, 22–23
Supplementary angles, 144–145
Surface area
    of cone, 194–195
    of cylinder, 194–195
    of prism, 192–193
    of pyramid, 192–193
    of sphere, 201
Symbols
    absolute value bars (| |), 12–13
    approximately equal to (≈), 110–111
    brackets, order of operations and, 2–3
    congruent figures (≅), 162–163
    decimal point (.), xvi, 120–121
    Egyptian, 141

greater than (>), 268–269
greater than or equal to (≥), 268–269
less than (<), 268–269
less than or equal to (≤), 268–269
not equal to (≠), 268–269
parentheses, order of operations and, 2–3
pi (π), 182–184
of right angle (⌐), 144–145
similar figures (~), 100–102
Symmetry, 168–169
Systematic random sample, 204–205

T

Tangent ratio, 110–111, 112–113
Tendency, central, 208–209
Terminating decimals, 34–35
Terms, combining like, 260–261
Theoretical probability, 242–243
Translations, 164–166
Transversal, 146–148
Trapezoid, 158–159
    area of, 178–180
Tree diagram, 234–235
Triangles, 160–161
    acute, 158–159
    area of, 178–180
    congruent, 162–163
    equilateral, 158–159
    isosceles, 158–159
    obtuse, 158–159
    Pascal's, 240–241
    right, 158–159
    scalene, 158–159
    similar, 100–102
    sum of angles in, 144–145
Triangular numbers, 3
Triangular prism, 188–189
Triples, Pythagorean, 77
Truth value, 83
Two-step equations, 24–25, 258–259

U

Unbiased sample, 204–205
Unfair game, 255
Unit rate, 86–87

V

Variables, 4–5
    on both sides of equation, 264–265
Variation, measures of, 212–213
Vertical angles, 144–145
Vertex, 144–145
Volume, 188–189
    of cone, 190–191
    of cylinder, 188–189
    of prism, 188–189
    of pyramid, 190–191
    of sphere, 201
Voluntary response sample, 204–205

X

x-axis, xxiii, 286–287
x-coordinate, xxiii
x-intercept, 286–287

Y

y-axis, xxiii, 286–287
y-coordinate, xxiii
y-intercept, 286

Z

Zero (0)
    addition of, 8–9
    as exponent, 54–55